EVOLUTION:

Reconciling
The
Controversy

John R. Hadd

Published by KRONOS Press in association with the *Center for
Interdisciplinary Studies,* Glassboro State College,
Glassboro, New Jersey, 1979.

Published by KRONOS Press

EVOLUTION:
RECONCILING THE CONTROVERSY

TABLE OF CONTENTS

To Judith

PREFACE

The goal of this work is stated simply: to attempt to bring man to his senses; for once that is accomplished, it is reasonable to hope that he could avoid what threatens to be a quite disastrous future.

To put the matter another way, the objective of the following essay is to bring about a restitution of perspective. If this can be done in timely fashion, then the perception of Ronald Higgins, that "Mankind is blundering headlong towards multiple calamity,"* need not become manifest destiny.

Higgins' is the latest warning. He describes six immense impersonal threats to the human future: nuclear abuse, overpopulation, famine, resource shortage, environmental degradation, and technologies racing out of control. But it is the human factor — the Seventh Enemy — that is the most critical part of the overall problem, the human factor as exemplified by the inertia of political institutions and obstinate individual blindness to the realities of our time.

Before effective common action in response to the six interrelated threats will be possible, a common understanding, both of man's situation and of his obligations in response to it, is required. There is no way in which the stewardship responsibilities of man can be reconciled with the dominant behavioral practices of the present era. To help achieve the requisite common understanding, and to clarify the nature of man's stewardship obligations, this work is devoted. It is directed explicitly at the issue — evolution — which has done, and is doing, the most to prevent common understanding and urgently re-

---

* Ronald Higgins, *The Seventh Enemy: The Human Factor In The Global Crisis* (New York: McGraw-Hill Book Company, 1978), p. 11.

quired collective remedial action.

The formulations contained herein originate in the West and is the first of many initiatives which must be undertaken if calamity of global proportions is to be avoided. But the audience which eventually must be reached is located in both East and West. I pray, therefore, that the message of this essay will be received in all quarters in the spirit of global family with which it is conveyed.

JRH

Arlington, Virginia
December 1978

Chapter I

THE PRIORITY TOPIC

Chapter I

THE PRIORITY TOPIC

The human family is in trouble now, and prospects for its future are increasingly grim. The potential for nuclear immolation grows insistently. Biomedical technology, which could emasculate humane life, mushrooms. Little or no effective response is being accorded to the deterioration of the globe's life-sustaining ecosystems.

These are but three examples of the fact that the human affairs of Spaceship Earth are very nearly out of both conceptual and practical control. The human family is, collectively, at the mercy of a grievously deformed perspective of life. Life is being worked out solely according to the script of secular man. As a consequence, it is being worked out badly and a lethal curtain is in prospect.

A sharply revamped script, a new set of governing ideas must be agreed upon. Without them, the family is virtually guaranteed to drift into catastrophic confrontation between both its members and the natural environment. The set of governing ideas — that which constitutes the controlling world/life paradigm — must be overhauled, overhauled to conform with the totality of knowledge and experience.

The experience of human life lends itself to a variety of interpretations. Because such diversity of perception adds to the richness of our lives and understanding, we should encourage every legitimate degree of diversity. But experience has also taught us that diversity of perception, taking the form of paradigm competition and conflict, can lead to the severest of dangers. Paradigm overhaul, leading to a new consensus on measures to protect human life is, consequently, a project of utmost priority.

The most obvious illustration of paradigm conflict, leading direct-ly to a grave threat to present and future human life, is the runaway nuclear arms race. Decisions and events of recent times belie the pro-nounced goal of both America and Russia for nuclear weapons con-trol and disarmament. Strategic Arms Limitation Treaty (SALT) negotiations move at a snail's pace, if at all. Even if a SALT II agree-ment is negotiated, prospects for its ratification in the Senate are heavily dimmed by "linked" factors such as the unremitting expan-sion of Soviet strategic forces and paramilitary maneuvering, as on the Horn of Africa. Concurrently, U. S. arms budgets and produc-tion continue growing: deferral of the B-1 bomber system, for exam-ple, only leads to shifting emphasis and funds to other options such as cruise missiles and neutron warheads. It is a horrid, open secret that an "Arms Treaty Is Promoted As A-Stockpile Expands":

> Leaders of the American and Soviet governments — whatever their differences — profess to have one common political priority this year: the conclusion of an agreement to limit strategic nuclear weapons. Yet the United States is today launching its most ambitious nuclear weapons pro-duction program in two decades to meet what its in-telligence agencies say are growing Soviet nuclear pro-grams.[1]

Both sides continue to be trapped mentally in an arms race without a foreseeable end — except for the prospective finis of nuclear blowup, accidental or otherwise.

Weaponry accident has been the most hushed aspect of nuclear arms stocking. The public has been deceived into believing that "fail-safe" control systems are functioning. "Fail-safe" propaganda masks the fact that the weaponry accident rate is presently calculated at *1 every 90 days*[2] and the risks worsen with every (daily) addition to weapons stockpiles. But one accident, in the "wrong" place, and Ar-maggedon could be upon the family. The recent fall of Cosmos 954, a nuclear-powered Russian satellite, is the latest illustration that the presence of "fail-safe" nuclear technology is an extraordinarily dangerous fallacy.

---

1. "Arms Treaty Is Promoted As A-Stockpile Expands," *The Washington Post,* November 19, 1978.

2. Lloyd J. Dumas, "National Insecurity in the Nuclear Age," *Bulletin of the Atomic Scientists,* May, 1976. Also, Stockholm International Peace Research Institute, *Year-book,* 1977.

Stopping the nuclear arms race and eliminating weapons stockpiles is, however, going to take something more than monotonous rhetoric and erratic negotiations between Russia and the United States. Even with 30-plus years of concern over the horrific consequences of nuclear trends, the problem has never received other than superficial analysis. We can no longer delay going beneath the surface and seriously examining basic issues.

To do so we must start with the question: why is there an arms race at all? The first answer becomes apparent immediately. The potentially lethal competition is a direct result of world/life paradigm conflict, of systems of interpretive ideas regarding the world — its origin, man's place in it, and life generally — in harsh conflict at many points. We must identify the points at issue and most responsible for perpetuating armaments competition and deal with them in a new consensus, if the actual cause of the nuclear arms race is ever to be effectively arrested.

The insane arms competition is but one of many socially enervating results stemming from the fact that the present world/life paradigm is, for the majority, dominated by the fundamental ideas of 19th century materialistic scientism. This point was made abundantly clear by E. F. Schumacher in his world best-seller, *Small Is Beautiful: Economics As If People Mattered.*[3] In that book, Schumacher, a highly respected scientist and teacher, identifies the "leading ideas" of materialistic scientism. They are: Darwinian concepts of evolution and natural selection; Marxian denigration of non-material aspects of human life; Freudian emphasis upon the "dark" subconscious of the human mind; the general idea of relativism, denying all absolute norms, moral or other; and philosophical positivism, which accords respect to the knowledge verification methods of only the natural sciences.

Schumacher notes that the foregoing set of ideas

> " . . . dominate, as far as I can see, the minds of 'educated' people today . . . No one, I think, will be disposed to deny the sweep and power of these six 'large ideas.' They are not the results of any narrow empiricism. No amount of factual inquiry could have verified any one of them. They represent tremendous leaps of the imagination into the unknown and unknowable. Of course, the leap is taken from a small platform of observed facts. These ideas could

3. E. F. Schumacher, *Small Is Beautiful: Economics As If People Mattered* (New York: Harper and Row, 1973).

not have lodged themselves as firmly in men's minds, as they have done, if they did not contain important elements of truth. But their essential character is their claim of universality. Evolution takes everything into its stride, not only material phenomena from *nebulae* to *Homo sapiens* but also all mental phenomena, such as religion or language. Competition, natural selection, and the survival of the fittest are not presented as one set of observations among others, but as universal laws. Marx does not say that some parts of history are made up of class struggles; no, 'scientific materialism', not very scientifically, extends this partial observation to nothing less than the whole of 'the history of all hitherto existing society'. Freud, again, is not content to report a number of clinical observations but offers a universal theory of human motivation, asserting, for instance, that all religion is nothing but an obsessional neurosis. Relativism and positivism, of course, are purely metaphysical doctrines with the peculiar and ironical distinction that they deny the validity of all metaphysics, including themselves.

What do these six 'large' ideas have in common, besides their non-empirical, metaphysical nature? They all assert that what had previously been taken to be something of a higher order is really 'nothing but' a more subtle manifestation of the 'lower' — unless, indeed, the very distinction between higher and lower is denied. Thus man, like the rest of the universe, is really nothing but an accidental collocation of atoms. The difference between a man and a stone is little more than a deceptive appearance. Man's highest cultural achievements are nothing but disguised economic greed or the outflow of sexual frustrations. In any case, it is meaningless to say that man should aim at the 'higher' rather that the 'lower' because no intelligible meaning can be attached to purely subjective notions like 'higher' or 'lower', while the word 'should' is just a sign of authoritarian megalomania."[4]

Paradigm overhaul aims explicitly at a valid alteration or moderation of the leading ideas of materialistic scientism so as to eliminate any cause for nuclear conflict, any rationale for continuing armaments competition. But one of the most useful aspects of

---

4. *Ibid*, pp. 82-3.

Schumacher's portrayal is the clear showing that the work to be done does not involve mere intellectual jousting with some "foreign" adversary; the leading/deforming ideas, and their influence, must be dealt with in every nation, without exception. Further, "overhaul" does not imply a blanket rejection of Darwin, Marx, Freud, or any other contributor to the presently dominant paradigm. Instead, the challenge is to establish which of their ideas and interpretations remain valid in light of accumulated evidence and the requirements of a comprehensive paradigm. What fits, stays. What does not fit goes into the doctrinal trashcan — regardless of the intellectual anguish its disposal may cause.

While Schumacher's scientific forté was developmental economics and appropriately-related technology, his most fervent conviction pertained to the need to break the world-wide conceptual stranglehold of undiscriminating materialistic scientism. As long as the leading ideas of materialistic scientism, the ones that Schumacher came to understand so well, are uncritically retained in our intellectual baggage, the human family is saddled "with a view of the world as a wasteland in which there is no meaning or purpose, in which man's consciousness is an unfortunate cosmic accident, in which anguish and despair are the only final realities."[5]

We need not, we must not, keep the human family subject to a paradigm where DESPAIR serves as the password of the "enlightened". Schumacher's two most important works, *Small Is Beautiful* and the philosophically-related *A Guide For The Perplexed*[6], constitute major contributions to the case against despair, a case others must now complete. Through *Small* and *Guide*, he initiated the great work of overhauling the world/life paradigm. This essay is meant to add to that initiative.

A comprehensive paradigm is required, one adequate to the demands of an interdependent global society, and likewise sensitive to all things important to a *mature* experience of human life. The latter requirement means that the place and practice of religion be fully and consciously provided for. In the ultimate conclusion of a career of intensive intellectual screening, Schumacher, born and raised in an atmosphere of traditional German skepticism, stated forthrightly that "It may conceivably be possible to live without churches; but it is not possible to live without religion. The modern experiment to do otherwise has proved a failure." He thereby reaffirmed the wisdom

5. *Ibid.,* p. 84.

6. E. F. Schumacher, *A Guide For The Perplexed* (New York: Harper and Row, 1977).

of the ages, and the overhauled paradigm must likewise reflect such wisdom. *Responsible* religion must be accorded a prominent place in a properly revised scheme-of-things.

I will comment further on religious aspects of the paradigm later in this essay but here wish to stress — as much for attention in the West as the East — two points. First, the pretentious case (extreme hubris) of "scientific atheism" has already been demolished.[7] Further criticism of it in this essay constitutes a necessary bombing of the rubble. Second, while I will urge, for plainly stated reasons, a Christian orientation for the overhauled paradigm, I do not hold that *only* Christianity can contribute meaningfully to it. All the major religions can, *must*, contribute to insure paradigm adequacy.

The largest single obstacle to restoring religion to respectability is traditional evolutionist doctrine. Martin Ling notes: "There can be little doubt that in the modern world more cases of loss of religious faith are to be traced to the theory of evolution as their immediate cause than to anything else."[8] If Ling is correct, why? Because, it seems certain, the "facts" of evolution have been presented in a manner which has " . . . imprisoned modern man in what looks like an irreconcilable conflict between 'science' and 'religion'."[9] Given the need for self-respect, given the amount of peer-pressure in a science-dominated era, given the absence of a credible alternative interpretation of geophysical events — particularly from the Church — people are inclined to adopt the "scientific" option, accepting as unavoidable the mental and spiritual discomfiture entailed in their choice.

As Schumacher acknowledged, the traditional evolutionary argument contains important elements of truth. The traditionalists are absolutely right in contending that the "message of the rocks" warrants keen appreciation. If biblical fundamentalists really desire to deepen our understanding of the Creation in order to enhance its prospects for survival, they must own up to the fact that conditions on Earth in the remote past were vastly different from what they are

---

7. Karl Marx fathered the notion of "scientific atheism". My book, *The Direct Connection* (see note #16 below), lays bare its fallacious nature. Worth particular note is the fact that Marx never engaged in theologic argument; he much more sought to bypass religion than try to refute it. Questions of the truth or falseness of religious dogmas were never a focal point of his concerns. He rejected religion because it was incompatible with his theory of desirable social action and personal freedom. He declared for atheism to try and instill a motivation for social change. He sought to banish God for the betterment of man. Rarely, if ever, has history witnessed a larger example of good intentions backfiring.

8. Martin Ling, cited in *Guide,* p. 115.

9. Schumacher, *Guide,* p. 114.

now (and have been through recorded historical time). Evolution — change, often dramatic — *has* occurred. An incredible amount transpired prior to "events" in the allegorical Garden of Eden, and further evolutionary change has occurred since they were depicted.

The traditional evolutionists do have some major evidence to support their interpretation. But, while they have selected points, they emphatically lack a case, especially one which could be justified as "scientific". Schumacher's characterization is totally apt: "Evolution *as currently presented* has no basis in science . . . (It) is not science: it is science fiction, even a kind of hoax . . . that has succeeded too well."[10]

We must deal, at some early time, with all of the major concepts that account for existing deformations in the world/life paradigm. In this work, however, concentration focuses on traditional evolutionary doctrine. That doctrine is selected as the "priority topic" for it appears to exert more foul and more perfidious influence than any other notion of materialistic scientism. Its influence has to be countered and offset, not just for the benefit of valid religion, but valid human thought and perception in general.

The next section of this essay will examine the six principle tenets of traditional evolutionary doctrine in order to show their implausibility and to present what I consider a philosophically and scientifically defensible explanation of evolution. The lessons learned in the second section set the stage, provide the framework, for initiating an effective response to the two problems of concern in the third and closing section: the control of biomedicine and the protection of Spaceship Earth's life-supporting ecosystems.

An effectively overhauled world/life paradigm is imperative for the protection of existing and future human life. Achievement of this goal will demand one major attribute: candor. We will examine, in part, the limits of authentic science while at the same time calling for support in the exposure of false science. *If this essay has any marked audience it is to all global subscribers to honest science, to those able to understand and willing to help clarify the enormity of what is at stake.*

---

10. *Ibid.,* (emphasis added).

Chapter II

THE RISE OF A CREED

Chapter II

THE RISE OF A CREED

Paleontology, the branch of geology that deals with forms of life through the study of plant and animal fossils, is reported "in an uproar, in ferment".[11] Relatively recent finds in different parts of the world have generated new speculation about the origin — and change — of life on Earth. Similar new finds may also pertain to the life form of man, *Homo sapiens.*

Much "convention" is now subject to fundamental reassessment. For example, leading sociobiologist Edward O. Wilson, has recently stated that the "natural selection theory does have severe structural weaknesses and the theory of evolution may be due for some important changes in the near future";[12] and Kenneth E. Boulding has already developed parallel ideas in his book *Ecodynamics; A New Theory of Societal Evolution.*[13] So I am not alone in believing that the present intellectual climate affords a highly opportune time for setting forth a substantive alternative to traditionalist evolutionary doctrine.

I therefore intend to propose such an alternative. It will accommodate that which is valid from Wilson, Boulding, or any other party. But it will go "beyond" any alternative so far developed while, at

---

11. Dr. J. William Schopf, paleontologist, University of California at Los Angeles, as cited in *The Washington Post,* October 3, 1977. The point of conceptual flux within the discipline is reinforced by paleonanthropologist David Pilbeam in "Rearranging Our Family Tree," *Human Nature,* June, 1978.

12. Edward O. Wilson, as cited by Tom Bethell in "Burning Darwin To Save Marx," *Harper's* Magazine, December, 1978, p. 92.

13. Kenneth E. Boulding, *Ecodynamics: A New Theory Of Societal Evolution* (Beverly Hills: Sage Publications, 1978).

the same time, "going back". I will set forth a framework within which it should prove possible to end over a century of acrimonious argument between evolutionist and creationist schools of thought and belief. It will be argued that there is no essential conflict between the Bible and evolutionary facts-of-life. Initially, such a contention is bound to appear as scientific heresy, even to evolutionary revisionists like Wilson and Boulding. The attitude to be anticipated has been summarized by Mortimer J. Adler in his valuable work, *The Difference Of Man And The Difference It Makes:*

> If the relative truth of the immaterialist hypothesis is confirmed, the problem of man's origin — both the origin of the race and the origin of the individual — may call for reconciliation of evolutionary theory with orthodox Christian theology. It is hardly an overstatement to say that most scientists today are unprepared for this eventuality.[14]

This prospect is not altogether discouraging, however, for most scientists have not, I believe, been properly exposed to a balanced examination of the issues. All that is asked of the readership at this time is an open mind and a willingness not to prejudge the matter. In the following pages, I hope to demonstrate in a fully satisfactory way that the Word and works of God, erudite theology, legitimate philosophy, and honest science are fully compatible.

I am mindful of the need for great care in dealing with the subject of evolution. The integrity of both science and Christian theology are at stake. At this stage in the essay, the emphasis is on the former. A notation of Richard E. Leakey, probably the most prominent anthropologist of the day, is in order:

> " . . . the Piltdown forgery illustrates the sometimes indecent eagerness with which scientists will accept what they want to believe. Researchers today are not exempt from this weakness, and it can still be seen in all branches of science. But because theories in archeology are often constructed from relatively little data, in that field the danger of over-interpretation, and therefore biased theories, is particularly acute."[15]

Speculations that the world confronting man was not an entirely

---

14. Mortimer J. Adler, *The Difference Of Man And The Difference It Makes* (New York: Holt, Rinehart and Winston, 1967), p. 291.

15. Richard E. Leakey and Roger Lewin, *Origins: What New Discoveries Reveal About the Emergence of our Species and its Possible Future* (New York: E. P. Dutton, 1977), p. 33.

set piece, that types of change were occurring in both the physical environment and to the life forms occupying it, were recorded as early as ancient Greece. The early Church fathers regarded such speculations as dire heresy, however, and sought to restrain further investigation of evolutionary possibilities. Even so, the subject continued to whet the intellectual appetite of many. Relevant data accumulated over time; and technological apparatus, such as the microscope, came to the aid of interpretation.

Eventually, the evidence against a one-shot, totally immutable, Creation became so great that men of discernment could no longer stay silent.

Charles Darwin spoke up, in 1859, through publication of *The Origin of Species,* though the act was taken with some reluctance and its timing influenced by the prospect of a similar exposition being published by Alfred R. Wallace. Darwin, once a devout Christian, had many years before admitted that the thought of making public his views on evolution saddened him and would cause him to feel as if he were "confessing a murder".

Darwin's theory was built upon three main sources: preceding conceptual work by Charles Lyell and Thomas Malthus, plus his own field observations of plant and animal species. The influence of Lyell and Malthus on Darwin cannot be overemphasized. Lyell was a staunch, articulate defender of geologic uniformitarianism and held that whatever geologic natural phenomena was under consideration, the present was like the past and that nature never exhibited drastic, sudden change. Darwin extended Lyell's view, regarding change in the inorganic world, to the world of living things. Darwin coupled that extension to another extension, relating uniformitarianism to Malthus' contentions regarding the struggle for existence among human populations. Darwin judged that Malthusian concepts gave insight into how various plants and animals survived, and pronounced his belief that favorable variations ("the fittest") are preserved solely through the neutral process of natural selection. Evolution, therefore, came to be explained basically as the natural selection of slowly changing forms of life.

Origin, which has often been attributed to have caused one of the greatest philosophical and intellectual revolutions of all time, has become an indelible part of history. It, coupled with Darwin's later publication of *The Descent of Man,* gave birth to traditionalist evolutionary theory, which itself has come to be built upon six main tenets. These tenets carry the general notions of uniformitarianism and the sameness of all matter to their absolute limit. The tenets are:

—the physical universe developed from unorganized matter, matter which has always existed;

—inanimate nature was responsible, through spontaneous chemical action, for the origin of life;

—existing animals and plants, reflecting unlimited mutability of species, developed by a process of gradual, continuous change from previously existing forms, starting with the simplest of one-celled ancestors;

—forces of nature, enforcing the principle of natural selection, are solely responsible for determining why some variations of life have survived and others have been lost;

—man himself is a product of the evolutionary process, and only that;

—there is no discernible purpose to or meaning in life, for all evolutionary change is neutral and without meaningful direction.

Before moving to an examination of traditionalist evolutionary doctrine through scrutiny of its main tenets, an apologetic word is in order on behalf of Darwin. He was a conscientious, contemplative man of scientific disposition who was disturbed by contradictions between fundamentalist teachings and observable facts. Darwin's conceptualizing aimed at trying to improve man's understanding of the workings of life, and his pioneering effort was an honest one; he can hardly be blamed for the actions of others in promoting what will be demonstrated to be a scientifically-deformed evolutionary "Creed".

As certainly as I will not accept or adhere to that creed, so would Darwin, I am sure, reject it. It would be a rejection founded on fact, yet neither of us would renounce belief in the occurrence of types of evolutionary change. Darwin tried to interpret, as best he could, with the evidence at his disposal; let us now do the same, with the evidence at our disposal. Let us act with a common, mind-instilled objective: to better understand.

I began refutation of traditionalist evolutionary doctrine in *The Direct Connection*.[16] Here follows a repeat of that section which addresses the evolutionary dispute. It originally appeared in Chapter IV (which analyzed "The Objections" to God and the Christian

---

16. John R. Hadd, *The Direct Connection: Rescue From Onrushing Global Catastrophe* (New York: Vantage Press, 1977), pp. 93-101.

religion) and, specifically, in that part of the chapter which deflates the long-standing assumption that conflict between science and Christianity is inherent and irreconcilable.

The most regularly and highly publicized conflict between science and the Christian religion stems from dispute over man's origin and his place in the natural scheme of things. Therefore, let's examine closely the subject of evolution and the Living Creator.

What is man's relationship to his historical and surrounding environment? What are the most notable features of the physical relationship? What are the principal implications between available evidence and the God of the Bible?

Questions of that sort have, since Messrs. Darwin and Huxley published, caused a savage battle, an endless, emotion-fraught fight between literal interpreters of the (Old Testament) Word and narrow-minded interpreters of physical findings supporting the theory of evolution.

Given the evidence now at hand, we see there is no need or rational basis for perpetuating the argument. Both sides have valid points. *There must be a reciprocal recognition.* As with the issue of free will and determinism, we must get away from analysis in terms of either/or.

Proponents of the "special creation" argument must end their lofty self-conceit about the posture of man in the scheme of things and quit blinking at solid evidence. Conversely, we must remove the blinders from advocates of the "nothing essentially different about man" position, for they are choosing to see only partial reality and are invoking on their behalf only selected portions of the evidence.

The allegory of the Creation, as set forth in *Genesis,* is as great a poetic achievement as man has produced. Perhaps it can be improved upon. The challenge is there for anyone intelligent enough to accept it and gifted enough actually to better human communication about what was and is involved.

The opening of *Genesis* is, however, a poetic achievement. It should not, it could not, and it never intended to serve as an event-by-event account of the physical evolutionary process. Stated simply and finally, there is no justification for invoking the narrative of *Genesis* as a source superior to geologic remains and other evidence indicating the relationship of man to other and lower orders of life.

A pattern is discernible. Man does show some physical resemblance to lower orders. Man is constrained by a number of biologic factors, in the same fashion as the lower orders. Man's organic structure is comparable to theirs in many important respects. These resemblances cannot be denied. The discernment of these resemblances is not something to condemn, just the opposite. We should be grateful that they have been found out. Given the nature of man's learning faculties; given the limits of his reason; given his dependence upon observation either to learn in the first instance or to confirm his mental deductions; then we can only value the fact that there is a discernible relationship between man and the lower orders. Given what man needs to know in order to survive as a species; given the tools he has for learning it; it would have been nothing other than a malicious trick on the part of God if he had set man off *completely* separate, distinct, and apart.

The time has come for biblical fundamentalists to become more alert. Likewise, the time to obliterate the absurd argument that man is nothing more than an animal — a mere naked ape — has also arrived.

From the first formulation of the theory of evolution, there have been acknowledged missing links in the chain of relationships between the major components of the Creation, the links which would give explanation between animate and inanimate, and also explain the differences between man and the other orders. The missing links are still that — missing — and we need to keep that fact constantly and prominently in mind.

At this point, before getting to the specific ways in which man's uniqueness is demonstrable, a short deviation is necessary. Our concern is with another traditional assumption about God, one similar to the old-cold-disciplinarian assumption, and one which now appears equally erroneous.

The assumption now challenged is that whatever Creation involved, once done it was finished. We have been prone to believe that God knew at the start what He wanted, and did it, both on earth and in heaven, and that it has all been finally established, all set up.

We cannot accept that assumption, for it is difficult if not impossible to reconcile it with either the physical evidence before us or the word of Jesus.

We believe in God the Creator, but not that His was a one-shot, finished-at-the-start, creation. We believe in a Living God who is continuously thinking and deciding and acting upon the full content of creation, including after-earth life. When the discussion moves to prospects of and in Heaven . . . we will consider carefully the advice of Jesus. Right now, we concentrate on relevant physical evidence.

When we are born, our brain size, about 130 cubic centimeters, is only slightly larger than that of a gorilla baby. This is why human and anthropoid young look so appealingly similar in their earliest infancy. A little later, an amazing development takes place in the human offspring. In the first year of life its brain trebles in size. It is this peculiar leap, unlike anything else we know in the animal world, which gives to man his uniquely human qualities ... If we accept the evidence of evolution, we must assume that man became by degrees, that he emerged out of the animal world by the slow accumulation of human characters over long ages—save for that seemingly rapid spurt in brain growth, which has carried him so far from his other relatives.[17]

It is submitted that this crucial difference between man and the lower orders, the establishment of conditions which permit *the totally atypical human brain growth,* is attributable to a conscious act of God. It is not necessary to know precisely how and when to understand that at a point in evolutionary time God made adjustments as were necessary to insure that His Creation would result in the procreation of intelligent, perceptive offspring. Even then His work was not done, according to Jesus, but that point is considered later.

Man *is* a special part of the Creation — man in His image, as a spirit; man, sharing His attributes of free will and intelligence, thanks to the brain which has been specially provided. Singularly or in combination their importance and presence is confirmed daily:

—Man alone is able to employ verbal speech and other forms of advanced communication. These allow him to compile a record of experience and pass on historical lessons so that each generation is not automatically condemned to a dull or gruesome, repetitive cycle of experience.

—Man is the least dependent upon fixed instincts. When man forsakes use of his intelligence and willpower; when he fails to use those gifts to temper and control his impulses; when he simply gives in to excitable instinct—as toward capricious sex, or violence—then he is not deserving to be classed as man. He forsakes legitimate status as a member of *Homo sapiens;* he voluntarily associates with the lower orders. "The uniqueness of humankind comes indeed from its potential ability to escape from the tyranny of its biological heritage. In-

17. Loren Eiseley, *The Immense Journey* (New York: Random House, 1957), pp. 109, 105.

stead of being slaves to their genes and hormones, as animals are, human beings have the kind of freedom which comes from possessing free will and moral judgment. We can repress our animality if we will."[18]

—Man alone demonstrates a craving to understand, to learn the purpose and structure of the Creation. He is the sole species independently capable of abstract calculations and conceptual learning. The result is that he not only has present knowledge, but can anticipate, especially anticipate death, and it is the capability for anticipation of death which gives him awareness of the value of life.

—Only man has the capacity to perform an ethical evaluation of his performance. Of all losses man has sustained in the past two hundred (Enlightened) years, "no deprivation has been so terrible as the abandonment of private guilt (for faulty performance). It was dreadful for him to lose a Creator; it was worse than dreadful, it was shattering for him to cast off responsibility. . . . . When society substituted shame for guilt, it amputated half of the human psyche, for it made both transgressions and virtues involuntary. Animals can feel shame. Only man can know that he is guilty of sin."[19]

—Man alone is toolmaker and tool user of any consequence. His is the only order of being with a capacity for molding his environment—either to the detriment or enhancement—of the quality of his earthly life experience. Only he has the capability to impact upon significantly, actually to modify, his surroundings. Man does more than influence his environment. Within large discretionary limits he directs his own evolution. Every other animal is shaped by the environment. To a degree, man is too. But to a far larger degree man shapes his environment, and his destiny, as both a species and as an individual member of the species. This is possible not just because of his native intelligence, but because for man alone are motives, purposes, and choices an active element of life.

—Finally confirming the point of man's uniqueness is God's very Word, as exemplified in Matthew 10:29-31, Luke 12:6-7, and other similar direct comments by Jesus:

> Are not five sparrows sold for two farthings, and not
> one of them is forgotten before God? But even the very

---

18. René Dubos, "The Humanizing of Humans," *Saturday Review-World,* Dec. 14, 1974, p. 77.

19. Phyllis McGinley, in the introduction to C. S. Lewis', *The Screwtape Letters* (New York: Time Inc. Special Edition, 1963), p. xviii.

hairs of your head are all numbered. Fear not therefore: ye
are of more value than any number of sparrows.

Because we have His very specific word (in addition to other
evidence), we can with confidence reject the ultimate conclusions of
the mod-behaviorists. Men are not mere naked apes. Neither are they
the same class as rats or pigeons—as the mod-behaviorists would
have us believe. Such arguments have been exposed as badly defective, and to persist in them is equivalent to trying to saddle a pseudo-
science upon man.[20]

Next, a comment by C.S. Lewis on a widely held philosophical
view:

Ever since men were able to think, they have been
wondering what this universe really is and how it came to
be there. And, very roughly, [three] views have been held.
First, there is what is called the Materialist view. People
who take that view think that matter and space just happen
to exist, and always have existed, nobody knows why; and
that matter, behaving in certain fixed ways, has just happened, by sort of a fluke [a miracle far exceeding anything
recorded in the Bible] to produce creatures like ourselves
who think . . . .

The [second] view is the Religious view. According to it,
what is behind the universe is more like a mind than it is
like anything else we know. That is to say, it is conscious,
and has purposes, and prefers one thing to another . . . .

[The third is] the inbetween view called the Life-Force
philosophy, or Creative Evolution, or Emergent Evolution. The wittiest expositions of it come in the works of
Bernard Shaw, but the most profound ones in those of
Bergson. People who hold this view say that the small
variations by which life on this planet "evolved" from the
lowest forms to Man were not due to chance but to the
"striving" or "purposiveness" of a Life-Force. When people say this we must ask them whether by Life-Force they
mean something with a mind or not. If they do, then "a
mind bringing life into existence and leading it to perfec-
tion" is really a God, and their view is thus identical with
the Religious. If they do not, then what is the sense in say-

20. John W. Sutherland, "Beyond Behaviorism and Determinism," *Fields Within Fields*
(Winter 1973-4, Number 10), published by the World Institute Council.

ing that something without a mind "strives" or has "purposes"? This seems to me fatal to their view. One reason why many people find Creative Evolution so attractive is that it gives one much of the emotional comfort of believing in God and none of the less pleasant consequences. When you are feeling fit and the sun is shining and you do not want to believe that the whole universe is a mere mechanical dance of atoms, it is nice to be able to think of this great mysterious Force rolling on through the centuries and carrying you on its crest. If, on the other hand, you want to do something rather shabby, the Life-Force, being only a blind force, with no morals and no mind, will never interfere with you like that troublesome God we learned about when we were children. The Life-Force is a sort of tame God. You can switch it on when you want, but it will not bother you. All the thrills of religion and none of the cost. Is the Life-Force the greatest achievement of wishful thinking the world has yet seen?[21]

Life-Force. Wishful thinking. That says it all.

"Believing is an *essential* human function without which [man] finds it almost impossible to understand or to integrate his personality."[22] Directly related, ". . . what we need [is] a great vision of the sweep of mankind and an understanding that we are just in the middle of the story, that it is a story far from ended, and that we have to contribute our part to keep it moving upward and outward . . . ."[23]

The traditional theory of evolution, or traditional teaching of the theory, fails in vital respects. Three conspicuous examples: they do not pay proper homage to the special place of His children in the order of things; they are either devoid or contradictory when it comes to consideration of purpose; and they make no allowance for exceptional events. Each one of those fundamental deficiencies can, on the basis of the evidence before us, be remedied without doing any intellectual disservice to those aspects of traditional theory which have been borne out. Said differently, the essential needs cited in the preceding paragraph can be satisfied.

We can readily, and honestly, conceive a Creation set in motion by God. It has been and is evolving. Its objective [explained later] is the

21. C. S. Lewis, *Mere Christianity* (New York: MacMillan Paperbacks Edition, fourteenth printing, 1971), pp. 31-35.

22. Eugene C. Kennedy, *Believing* (New York: Doubleday and Co., 1974), p. 17 (emphasis added).

23. Theodore Hesburgh, as cited in Kennedy, *Believing*, p. 98.

fostering of intelligent human life—life which is both related and uniquely distinguishable from all other forms. He has provided for all essential human needs. There are no missing links of consequence for, as signified by Christmas, the *imperative* God—man link has been consciously forged. Unlike monkeys, men have the capabilities and responsibility to keep the Creation moving along its intended path . . .

The last general point to be considered in this section is allowance for the exceptional. Traditional theory contends evolutionary change has occurred and can only occur through minute, chance mutations, tiny changes occurring in undramatic fashion, accumulating over eons of time, happening to result in (among other things) man's embodying the highest intelligence. The argument regarding human brain growth as atypical would therefore appear as a claim for a giant "exception" to the "rules," and therefore be scientifically unallowable. Such dismissal might be tolerable save for the tremendous fact that there are veritable mountains of physical evidence, available for inspection by anyone willing to use his eyes, which absolutely cannot be accounted for under assumptions of minute, undramatic evolutionary change.[24] And the instant that evolutionary theory makes allowance for "one" exception, as it *must* in coping with the geologic record just cited, it is denied the theoretical basis

---

24. Immanuel Velikovsky, *Earth in Upheaval* (New York: 1955). Two points require mention. First, read the work and judge it yourself. How can the assembled evidence, the mountains of bones and stones and their obvious "dramatic change" implications, be denied? Second, be familiar with the fact that for a quarter-century Velikovsky has been the object of unremitting vilification by "the Establishment" since his works pose fundamental challenges to a number of established biases and assumptions in a number of disciplinary fields. Despite subsequent events having totally confirmed an impressive number of his "heretical" conclusions, "the Establishment" persists in denying his work a full and open trial on the merits. On February 25, 1974, the American Association for the Advancement of Science sponsored a symposium on "Velikovsky's Challenge to Science," ostensibly to show the world once and for all that his contentions were invalid. The proceedings provided only another chapter in this story of scientific swindle. See the report in *Pensée* (published by the Student Academic Freedom Forum, Portland, Ore.) Vol. 4, No. 2., Spring, 1974. Especially notable is the critique (pp. 38-40) of the major paper presented at the symposium, which aimed at "hard" analysis and disproof of Velikovsky's contentions about workings of the solar system. The "learned critic" won the day according to popular reports. Then the critique in *Pensée* carefully documented the fact that the "learned critic" based his counter-arguments on a system of astrophysics which makes *no* allowance for electric and electromagnetic forces. That is no minor omission, for as Geoffrey Chew notes: "Electromagnetism, together with space-time, is an essential fiber within the fabric that underlies the scientific description of nature." *Impasse for the Elementary Particle Concept,* in *The Great Ideas Today: 1974* (Chicago: Encyclopedia Britannica, Inc., 1974), p. 121. I will say it more directly: "Establishment" reliance on a "rebuttal" of Velikovsky which does not take account of what are recognized as elementary facts only means that a willful fraud continues. Either that is the case or the spokesmen for the AAAS have forfeited the right to judge anyone or anything. See also note 85 below.

for smug dismissal of "other" exceptions. If there is before us undeniable proof of one exception, what scientific basis have you for ruling out a second, or a third, or many? The only honest answer is that there is no such basis. The reader may not accept contentions regarding dramatic change, but he must do so with the understanding that traditional (Darwinian) theory cannot be invoked as the basis of rejection.

> The claim that chance mutations preserved by natural selection provide the *complete* explanation of the emergence of higher forms of life and its complex, purposive forms of behavior may turn out to be the swan song of a presumptuous generation of biologists. There is today a growing number of eminent biologists who have come to realize that chance mutations may provide part of the explanation, but not the whole explanation, and perhaps not even an important part of it.[25]

In summary, and to conclude this section, we state our confidence in the Living Creator, knowing perfectly well the work He is about. The facts of evolution and the evolutionary process are no cause for embarrassment, to Christians or anyone else.

Since the foregoing went to press I have encountered a number of works which further buttress the interpretations set out in *The Direct Connection* and which add strength to the alternative framework I offer to traditionalist evolutionary doctrine. These works will be introduced, and related, in the following serial examination of the six paramount tenets upon which the traditionalist creed is based. They implicate the entire cosmos. But it is not the range of their extent, the size of tenet coverage if you will, which is objectionable. To learn what is, we begin at the beginning.

ORIGIN OF THE UNIVERSE

Was there a beginning or was there not? As C.S. Lewis noted, the question has been debated for millennia. Hear Lucian, a Greek of the second century: "Some say the world had no beginning, and cannot end; others boldy talk of a creation . . . though it is by no means obvious how there could be a place or time before the universe came into being . . . Some circumscribe the All, others will have it

---

25. Arthur Koestler, "Order From Disorder," *Harpers,* July, 1974, p. 64 (emphasis in original).

unlimited.''

In our time, this debate has continued in the terminology of the ''steady-state'' versus the ''big bang'' alternatives. Proponents of the steady-state position assume that the universe had no moment of creation, that it has always been the way it is, perpetually expanding, and preserving its overall structure through continuous creation of gaseous hydrogen, created *out of nothing* in empty space. But this interpretation, according to the most advanced theory and accumulated evidence, is now considered by leading astrophysicists to be demonstrably inadequate. One of mankind's most ancient disputes appears to be reaching a resolution.

The explanation of this truly momentous and ''historical'' occurrence is conveniently provided by Steven Weinberg in *The First Three Minutes: A Modern View of the Origin of the Universe.*[26] In the mid-1960's astronomers made a discovery which tilted the interpretive scale decisively in favor of a ''big bang'' start. During that period it became understood that the universe is permeated with a microwave radiation which can hardly be explained unless one assumes it to be an electromagnetic glow remaining from a primeval explosion that constituted the birth of the universe. The presence of this microwave radiation accounts for the widespread preference for the big bang theory and serves as the basis for cosmic modeling efforts which integrate the laws of thermodynamics, relativity theory, and quantum mechanics.

Weinberg explains the workings of the big bang model in a very imaginative way, in the format of a motion picture which is stopped intermittently so that individual picture frames might be carefully examined. The first frame-stop is at one-hundredth of a second after zero time. The temperature of the universe is calculated to have been around 100,000 million degrees Kelvin. The universe, at that point, consisted of a soup of primitive particles and radiation, 4,000 million times denser than water and expanding at unimaginable speed. The fifth frame stops the action three minutes later. The soup has now cooled to approximately 1,000 million degrees (70 times as hot as the center of our Sun) and tritium and helium nuclei begin to take shape. The model indicates that it takes 700,000 years for electrons and nuclei to unite in stable atoms; then galaxies and stars begin to form. Roughly 10,000 million years after that, human beings attempt to reconstruct what transpired.

---

26. Steven Weinberg, *The First Three Minutes: A Modern View of the Origin of the Universe* (New York: Basic Books Inc., 1977).

The relation of the big bang interpretation to the first two laws of thermodynamics warrants special mention. The first law deals with the quantity of matter/energy in the universe. It holds the quantity to be constant: matter and energy within the universe can neither be created nor destroyed; and though they may change from one form to another $(E = mc^2)$, the total amount remains unchanged. According to the first law, the universe could not have created itself, to have come into being through spontaneous processes.

The second law of thermodynamics, the entropy principle, concerns the quality of energy—and its inexorable degeneration. At the inception, all energy was totally compacted and completely ordered. Over a period of time, that order deteriorated and energy becomes randomly spread. Eventually, maximum entropy and randomness will prevail. The universe will then experience "heat death". But the very direction of the universe—one way, running down as like a slowly unwinding clock—clearly infers that cosmic processes had a start fixed in time which initiated the direction.

Both the first and second laws, therefore, are in opposition to the first tenet of the traditional evolutionist creed. Further, the big bang interpretation is supported from the testimony of another vital source—that of the fossil record. The findings of the astrophysicists and the paleontologists square precisely, both affirming that there was *no* "life at the beginning" (a point to which we'll return later). The soup had to cool for a considerable period before the Earth proper could form and conditions accommodative to life could evolve. The age of the Earth is presently estimated to be 4.5 billion years. The oldest signs of life on earth—fossilized, microscopic one-celled creatures—are estimated to be 3.5 billion years old.[27]

A final consideration for this section: The best available evidence clearly indicates that the universe had a beginning. But was it the first? Asked differently, are we involved in and interpreting only an oscillation of an endlessly expanding-contracting universe? Dr. Arno A. Penzias, a co-discoverer of the microwave radiation radio signal — a feat which earned him one of the 1978 Nobel prizes — has been working on that very question. He finds that ". . . the data we have in hand right now clearly shows that there is not enough matter in the universe, not enough by a factor of three, for the universe to fall back."[28]

---

27. Throughout this book, age figures cited for fossils and the Earth itself are those now generally accepted and based on radioactive carbon and similar dating methods. While such estimates are subject to change, the principles set forth in the book are not dependent upon precise age calculations.

28. Dr. Arno A. Penzias, as cited in *The New York Times,* March 12, 1978.

Dr. Robert Jastrow, director of NASA's Goddard Institute for Space Studies, believes the evidence against an oscillating universe even stronger. In the course of examining the central implication of today's astronomical evidence ("Have Astronomers Found God?"), Jastrow states that ". . . the latest measurements indicate that the expansion of the universe will continue forever, because the amount of matter in the universe has turned out to be 10 times too little to exert the gravitational pull that would be needed to halt the outward movement of the galaxies . . . As usual when faced with trauma, the mind reacts by ignoring the implications — in science this is known as 'refusing to speculate' — or trivializing the origin of the world by calling it the big bang, as if the universe were a firecracker . . . A sound explanation may exist for the explosive birth of our universe; but if it does, science cannot find out what the explanation is . . . It is not a matter of another year, another decade of work, another measurement, or another theory. At this moment it seems as though science will never be able to raise the curtain on the mystery of Creation . . ."[29]

In summary: There was a beginning. It was an original event; and no scientifically or philosophically valid basis exists for assuming other than that an "outside," supernatural agency was involved with getting things going.

ORIGIN OF LIFE

Traditional evolutionists assume that inanimate nature was responsible, through spontaneous chemical action, for the origin of life on Earth. Their position is somewhat akin to the once scientifically accepted hypothesis that life could emerge spontaneously from decaying matter; for example, rotting banana skins often appear to generate fruit flies spontaneously.

In *Origin,* Darwin apparently disavowed belief in spontaneous generation. The closing paragraph of his book states: "There is grandeur in this view of life, with its several powers, having been originally breathed by the Creator into a few forms or into one . . ." While the disavowal remains in current versions of the text, its reten-

---

29. Robert Jastrow, "Have Astronomers Found God?" *The New York Times Magazine,* June 25, 1978, pp. 22, 29. See also Jastrow's book, *Until The Sun Dies* (New York: W. W. Norton and Co., 1977), Chapter 4, which concludes (p. 38) that "The latest answer is that all the matter and energy in the Universe, in these many forms, will not suffice to bring the expansion of the galaxies to a halt. According to the available facts, the materials of the Universe must disperse forever, until all is space and emptiness. It appears that there was only one beginning, and there will be only one end."

tion is misleading. Darwin's commitment to the theory of uniformity led him to say, in 1882: "Though no evidence worth anything has as yet, in my opinion, been advanced in favor of a living being being developed from inorganic matter, yet I cannot avoid believing the possibility of this will be proved someday in accordance with the law of continuity."[30]

Darwin was perhaps unaware of the fact that another scientist of his very generation, Louis Pasteur, had caused the final exploding of the theory of spontaneous generation. Pasteur completed the groundwork upon which has risen the law of biogenesis, which proscribes the spontaneous formation of life. Countless experiments have reaffirmed that if life-containing matter is sterilized and then kept from any other biological contact, no new life arises. The law also dictates that life will also only perpetuate its own kind and type. The term "law" as used here applies in the strongest scientific sense. Not a single exception has ever been observed or proved capable of inducement. That most crucial fact notwithstanding, the notion of spontaneous chemical origins continues to receive prominent promotion.[31]

No one, not even practitioners of contemporary biology, with its immensely powerful technologies (a subject of the final section of this essay), has been able to accomplish experimentally the formation of living matter from non-living matter. There has been no duplication through applied intelligence of what is assumed/claimed to have happened through the blind interaction of inorganic substances. This generalization covers all developments to date, including work with recombinant deoxyribonucleic acid (DNA).

Traditional evolutionists are well aware of the law of biogenesis. They choose, however, either to ignore it or, more commonly, downplay and evade the significance of its implications through an argumentative stunt of major proportions. The evolutionists stand their sacred principle of uniformitarianism on its head and now contend that past (primordial) conditions were so different that present processes and observations are irrelevant to the point at issue.

The inconsistent application of uniformitarianism is tolerated here, for we agree with the traditionalists that primordial conditions were vastly different. But even so, no grant of validity is given to the contention of spontaneous generation. The price of such a conces-

---

30. Charles Darwin, as cited by Robert T. Clark and James D. Bales, *Why Scientists Accept Evolution* (Grand Rapids, Michigan: Baker Book House, 1966), p. 45.

31. See, for example, Richard E. Dickerson, "Chemical Evolution And The Origin Of Life," *Scientific American,* September, 1978, pp. 70-86.

sion is exorbitant — the sacrifice of legitimate science.

That judgment is buttressed, in large measure, by a relatively new and so far virtually unknown tremendous assembly of fact, *The Creation-Evolution Controversy,* by R.L. Wysong.[32] It deserves rapid adoption as a standard textbook. All the major issues are laid out, and all in a fashion that allows the reader to judge independently which side has the most persuasive evidence and logic at its disposal. No issue receives more extended, or informative, treatment than the subject of the origin-of-life. Following is a sample of Wysong's work.

Proteins: In the absence of tissue and cellular protein, life is impossible. Wysong therefore begins his analysis of the origin-of-life issue with an explanation of protein formulation.

Proteins are built by/upon amino acids, 20 kinds, all but one of which can exist in two different forms, "L" or "D". All proteins derived from living organisms are composed of *only* "L" forms. All synthesized proteins (manmade chemical products) *always* entail a 50-50 D-L mixture. If life were mere chemical interaction, D-L protein constitution would be the norm. But, for actual life, L-only protein makeup is the norm.

Wysong explains how probabilities of chance formation of life must be calculated and how the laws of probability teach that "events whose probabilities are extremely small never occur". Non-occurrence, or impossibility, is ascribed by experts to events whose probability is in the range of or lower than $\frac{1}{10}^{30}$ to $\frac{1}{10}^{50}$. Wysong calculates the odds which would have to be overcome for just a single, pure L-amino acid protein to come into existence in a primordial chemical soup of mixed D-L forms. The answer:

$\frac{1}{10}^{114}$ That is one chance in one quintosexdecillion.

The odds against chance formation of life grow when it is recognized that useful proteins must, beyond containing only L-type amino acids, have a very specific sequence of those acids. No mere jumble of acids will suffice. The most basic living organism would require 124 proteins of properly sequenced amino acids. The odds of even the simplest living organism forming by chance are $\frac{1}{10}^{78,436}$.

---

32. R. L. Wysong, *The Creation-Evolution Controversy,* (Midland, Michigan: Inquiry Press, 4925 North Jefferson Avenue, zip 48640, 1976).

Origin and Modification of DNA: DNA provides the basis for life's unique reproductive capacity. All life, from bacteria to man, depends on DNA. It contains the instructional code which assures faithful replication of life forms. Wysong points out that, in the case of man for example, 3 billion cells in our bodies die every minute and are consistently replaced according to type due to the working of DNA.

Scientists have been able to synthesize the building blocks of DNA in the laboratory. But what is the potential that the nucleotides of DNA could arise by chance, arranged in proper code sequence, and then undergo spontaneous rearrangement thereby generating more and more complex organisms? Wysong stresses that DNA is not just another form of matter but is a code, a language carrying information. The volume of information contained in living organisms is phenomenal. Wysong notes, for example, that a simple one-celled bacterium, *E. coli.,* is estimated to contain information bits equivalent to 100 million pages of the *Encyclopedia Britannica.*

"... the creationist suggests that there is a fallacy in even suggesting biochemical bias as the mechanism originating life. He reminds us that life depends upon the sequence of repeating subunits. There are only four DNA letters. How could molecular preferences dictate various sequences of these four letters to form instructions for the myriads of different functions within organisms? If there were preferences, then only one word would repeat itself over and over. How could the repetition of one word yield the prodigious information found in even a simple clam? If bias produced a sequence, it would produce only one sequence, one monotonous simple organism. But life exists in tremendous complexity and abundant structural and functional variations.

The same argument applies to proteins which are made up of some 20 repeating subunits. If there were biochemical bias to the degree that evolutionists would suggest, then all life should be simple (although simplicity and life are opposites), and be similar to or the same. But it is not.

The sequence of nucleotides in DNA and the sequence of amino acids in proteins are sequences of language ... language that imparts meaningful information for the accomplishment of purposeful work."[33]

---

33. *Ibid.,* p. 125.

Communication science, Wysong next explains, has confirmed that information does not and cannot arise spontaneously. It only arises as a result of the expenditure of energy and the action of intelligence. The traditionalists attribute ascending evolutionary transformations to random mutations. Yet random changes in DNA symbol sequences would not enlarge information content; random change would invariably destroy meaning. And when modern computers are employed to simulate the assertion of DNA code "growth" through mutations and natural selection, the situation is found to be inconceivable — the computer jams.

DNA-Protein Totals: Life requires the simultaneous existence of DNA and protein. Proteins depend on DNA for their formation. But DNA cannot form and exist without pre-existing protein. Wysong calculates the chances of the interdependent DNA-protein life relationship evolving spontaneously. The total probability, even under highly liberal assumptions, of the chance formation of the proteins and DNA required by the smallest self-replicating entity is $\frac{1}{10}^{167,626}$.

Truly, as Wysong says, something has made the probabilities of chance creation of life "look moronic". Whatever is left, at this stage, of the traditionalist's argument is further weakened when close examination is given to prospects for original life surviving in a raw (chance), primitive environment. Wysong explains:

"If life is to have spontaneously evolved, each progressive step must be compatible with the conditions that brought it about. But we have seen that what little product could be formed is destroyed by the very agency that created it . . . .

A summary of explanations is interesting. It has been shown that life could not form in the presence of an atmosphere containing oxygen, so life must have formed in a reducing atmosphere. Yet that couldn't be, since living organisms need oxygen to survive, and without atmospheric oxygen there would not be an ozone canopy to shield the fragile chemicals of life from ultraviolet light. Therefore, life must have formed under the protective shield of several feet of water. But if the chemical precursors to life were under several feet of water, what would supply the chemicals with the necessary energy to react and form more complex molecules, and how could proteins be formed if water favors their dissociation rather than their synthesis? Furthermore, what would prevent currents within the water from bringing the chemical precursors to the surface where they would meet their demise from the action of the ultraviolet rays? It must be, therefore, that the biochemicals were formed on dry clays and rocks, but if that were so then we're back to the original problem of the chemicals being exposed either to the destructive powers of oxygen (oxidizing atmosphere) or ultraviolet light (reducing atmosphere) . . .

The materialist has trouble explaining the chance formation of life because life and its environment is a highly ordered integrated system. The present ecological crises testify to the highly tuned and integrated nature of the life-environment system. Just as it would be difficult to explain how a complex machine could operate with vital gears, circuits, pulleys, and levers missing or defective, so it is difficult to explain the piecemeal formation of the life-environment system on this planet." [34]

What is all too often taught about these matters is something terribly different, however, and is responsible for having roused Schumacher's deep ire in *Guide*. He cited *The New Encyclopedia Britannica* (1975) to illustrate how standardized traditionalist misrepresentations have become.

"The Doctrine of Evolutionism is generally presented in a manner which betrays and offends against all principles of scientific probity. It starts with the explanation of changes in living beings; then, without warning, it suddenly purports to explain not only the development of consciousness, self-awareness, language, and social institutions but also the origin of life itself. 'Evolution,' we are told, 'is accepted by all biologists and natural selection is recognised as its cause.' Since the origin of life is described as a 'major step in evolution,' we are asked to believe that inanimate matter is a masterful practitioner of natural selection. For the Doctrine of Evolutionism any possibility, no matter how remote, appears to be acceptable as if it were scientific proof that the thing actually happened.

When a sample atmosphere of hydrogen, water vapour, ammonia, and methane was subjected to electric discharges and ultraviolet light, larger numbers of organic compounds . . . were obtained by automatic synthesis.

---

34. *Ibid.,* pp. 216-18.

This proved that a prebiological synthesis of complex compounds was possible.

On this basis we are expected to believe that living beings suddenly made their appearance by pure chance and, having done so, were able to maintain themselves in the general chaos:

> It is not unreasonable to suppose that life originated in a watery 'soup' of prebiological organic compounds and that living organisms arose later by surrounding quantities of these compounds by membranes that made them into 'cells.' This is usually considered the starting point of organic ('Darwinian') evolution.

One can just see it, can't one: organic compounds getting together and surrounding themselves by membranes — nothing could be simpler for these clever compounds — and lo! there is the cell, and once the cell has been born there is nothing to stop the emergence of Shakespeare, although it will obviously take a bit of time. There is therefore no need to speak of miracles *or to admit any lack of knowledge.* It is one of the great paradoxes of our age that people claiming the proud title of 'scientist' dare to offer such undisciplined and reckless speculations as contributions to scientific knowledge, and *that they get away with it.* "[35]

The exposure and elimination of such false science, the teaching of which constitutes brainwashing and not education, is a requirement of highest priority. The problem will be addressed further as this essay proceeds. Here, however, we must concentrate on understanding the uniqueness of life.

Life is not a simple stage in an evolutionary continuum. An enormous gap exists between even a rich organic environment and the simplest organized life. Wysong explains what, in addition to DNA and protein, was required for the first living entity to become as well as the staggering complexity of just a single living cell. Life is organization, the contrary of entropic-oriented lifeless matter. Life is not a mere aggregate but an integrate. *Matter is a property of life but life is not just another property of matter.*

Wysong, synthesizing the conclusions of leading origin-of-life researchers, spells out the distinguishing characteristics of life and the best definition so far derived.

> "Life means information content, a definitive structure,

---

35. Schumacher, *Guide, op. cit.,* pp. 112-113 (emphasis in original).

the ability to seek out and consume food, expel wastes, metabolize, regulate activities and reproduce. Microspheres and coacervates perform none of these functions . . . .

Life, in the absence of interdependency and complexity is unheard of. Life rides upon complexity, not DNA alone, nor on protein, a membrane or organelles alone.

So then, what is life? Incorporating the ideas we have discussed: *Life is a code describing an urge for self maintenance and perpetuation in combination with the mechanism and machinery that allows the independent expression of that code* . . . .

On the other hand, anything less than life cannot sustain or perpetuate itself without the concomitant existence of life. Non-living matter may accrete, or 'grow,' such as is the case with crystals, or the building of polymers in the laboratory, but these processes will not continue beyond a point. There is no continuous functional drive to a higher level of complexity. The fate of anything less than life is degradation, dissolution, destruction and return to randomization. Non-life reaches a stalemate, an equilibrium, a static state: life has continual direction and progressive urge.''[36]

How did life originate? It came into being, per the Bible and massive supporting evidence, as the result of a *second creative act*. Just as surely as the traditionalist evolutionary hypothesis has never been proved, supernatural creation has never been disproved. His Word and the evidence are in complete agreement: evolution of the planet and its basic life forms have entailed a *series* of purposeful, creative acts. We will next see how the "message of the rocks" is a further affirmation of the creationist interpretation.

ANCESTRAL FORMS

Traditional evolutionists contend that existing animals and plants, reflecting unlimited mutability of species, developed by a process of gradual, continuous change from previously existing forms, starting with the simplest of one-celled ancestors. In other words, life forms are assumed to be in a ceaseless process of transformation, continuously changing in response to environmental pressures (forces of

36. Wysong, *Controversy, op, cit.,* pp. 229, 198 (emphasis in original).

nature).

In judging the validity of the unlimited mutability/transformation assumption, one type of evidence counts far more than any other — that of the fossil record. The content of that record refutes traditionalist claims and rejection of their assumptions is embodied in the title of the next work to be considered, *The Transformist Illusion,* by Douglas Dewar.[37] Dewar devoted over 40 years of professional life to studying the issues of evolution. *Illusion* was published in 1957, posthumously. It is partly out-of-date; for example, no pre-Cambrian fossils had been discovered before Dewar's death and he considered their absence a fatal flaw in the theory of evolution. That void in the paleontological record has since been filled and Dewar's objection rendered moot.

But Dewar's work is only partially dated. The majority of it is highly relevant to the issues at hand. Dewar's discussion pertains to the entire fossil record in contrast to, say, Leakey's *Origins* (cited above) which concentrates only on fossils considered germane to understanding the emergence of *Homo sapiens.*

Dewar demonstrates, over and over, *the true lessons to be learned from fossils are the direct opposite of those popularly taught.* The more the geologic record is honestly studied, the stronger becomes fossil refutation of traditionalist contentions. "The failure to recognize this fact is the reason why Transformism has not been relegated to the museum of exploded theories."[38]

Dewar sets forth four tests:

"If the evolution theory be true, the record of the fossils should exhibit the following features.

I. Every class, order, family, and genus would make its appearance in the form of a single species and exhibit no diversity until it has been in existence for a long time.

II. The flora and fauna at any given geological horizon would differ but slightly from those immediately above and below except on the rare occasions when the local climate suddenly changed if the sea flowed over the land, or the sea had retreated.

III. It should be possible to arrange chronological series

37. Douglas Dewar, *The Transformist Illusion* (Murfreesboro, Tennessee: Dehoff Publishers, 749 N.W. Broad Street, zip 37130, 1957).

38. *Ibid.,* p. 13.

of fossils showing, step by step, the origin of many of the classes and smaller groups of the animals and plants. By means of these fossil series it should be possible to draw up a pedigree accurately tracing the descent of most of the species now living from groups shown by the fossils to have been living in the Cambrian period.

IV. The earliest fossils of each new group would be difficult to distinguish from those of the group from which it evolved, and the distinguishing features of the new group from which it evolved would be poorly developed, e.g., the wings of birds or bats."[39]

The available evidence does not meet these tests. Before describing the actual evidence, it should be noted that a standard complaint of the traditionalists is that such tests are too stringent, that more is being demanded of the fossil record than is reasonable. Such arguments show the traditionalists trying to have things both ways. On the one hand, the paleontological record, giving clear evidence of extinct forms of life, is invoked as proof of the standard evolutionary hypothesis. On the other hand, when the fossil record is shown not to harmonize with traditionalist claims, the fault is said to be with an imperfect fossil record, not standard assumptions. Such argumentation smacks strongly of hypocrisy. Dewar cites data in an opening chapter giving some evidence that the fossil record is far more complete than traditionalists are willing to admit. Further, I would ask what of real consequence for human understanding is missing, since the fossil record provides evidence from the first one-celled species (what could be simpler, a more basic starting point for life) through all geologic periods?[40]

Rather than showing a traceable continuity, the fossil record of the various geological eras show profound change between them, and a stark absence of intermediate/transitional fossils. "As we pass from horizon to horizon we see much replacement of species and genera, and the first appearance of some families, orders, and classes, but no evidence that any genus, or larger group is the modified descendent of a group of which fossils occur at any earlier horizon . . . The

39. *Ibid.*, p. 35.

40. Boulding, for example, states: "We do not really know what produces rapid evolutionary change at one time and very slow change at another. We are trying to interpret a record so fantastically imperfect that the details elude us." (*Ecodynamics, op. cit.,* p. 326.) I submit the essential problem is not with the quality of the geologic record but with prior interpretations of its meaning.

abruptness with which new Classes and Orders of animals make their first appearance in the rocks known to us is one of the most striking features of the geological record."[41]

Dewar walks the reader through "The Succession of the Faunas": Cambrian, Ordovician, Silurian, Devonian, Carboniferous, Permian, Triassic, Jurassic, Cretaceous, Eocene; and, in a later chapter, Oligocene, Miocene, and Pliocene. The message of the rocks is consistent throughout: the geological data emphasize the sudden appearance of new groups, and extreme slowness with which change in animals has taken place.

"Future generations will comment on the fact that almost every adherent of the theory of organic evolution, when setting forth the evidence for the theory, assigns a minor place to the evidence of the fossils . . . The reason for this seemingly strange way of supporting Transformism is that the fossils are hostile witnesses. *Not a single fossil of vital importance for the support of the theory has come to light* . . . The fossil record shows that the earliest fossils of each class and order are not half-made or half-developed forms, but exhibit, fully developed, the characteristics of their class or order. Any changes undergone by the great group after it has appeared are comparatively insignificant. For examples, the pterodactyls, turtles, ichthyosaurs, bats, cetaceans, sirenia, seals, etc., when they first appear exhibit all the characters which distinguish their class or order and after that undergo hardly any change . . . The experiments of scientific breeders or geneticists tell the same story . . . Contrary to the expectations of transformists, breeding experiments have demonstrated the stability of species. [After much experimentation] Drosophilia melanogaster [fruit fly] still remains Drosophilia melanogaster. So it is with all the other animals on which geneticists and breeders have operated: the shrimp Gammarus, mice, guinea-pigs, rats, rabbits, pigeons, horses, etc. The proposition I submit . . . is the changes that have been effected gradually in animals are strictly limited and do not transgress the limits of the natural family . . . This is to assert, not that the family is the unit of creation, but that there is no fossil PROOF that any family is derived from an earlier one."[42]

I believe that Dewar's judgments continue to be true. Certainly the discovery of pre-Cambrian fossils does nothing to invalidate his

41. Dewar, *Illusion, op, cit.*, p. 36. Dewar is certainly not the only one to have noted the phenomenon of "abrupt appearance". For an excellent summary of other confirming works, see Lewis M. Greenberg, "Cataclysmic Evolution," KRONOS, Vol. I, No. 4 (Winter, 1976), pp. 98-110.

42. Dewar, *Ibid.*, pp. 13 (emphasis in original), 58, 150, 66.

general conclusions; neither do the most recent fossil findings by Leakey and others (discussed further below). There are some great lessons to be learned from his compendium of fossiliferous fact. The first, I would say, is further evidence on behalf of intermittent Creator involvement. Dewar concludes: "The rocks cry out creation!"[43] I fully agree and ask how to more reasonably account for the "abrupt appearance" of new classes and orders of animals?

The second lesson is that the fossil record serves as further validation for the law of biogenesis while further undermining the traditionalist assumption of unlimited mutability. There has been evolutionary change in life forms, but it has not been of either the magnitude or extent traditionally assumed. There is nothing wrong with students of evolution seeking to discern and understand the pattern of life's development. There is everything wrong with persisting in interpretive efforts through the traditionalist paradigm which is shot through with fault and contradiction. Its relation to the laws of thermodynamics is one example of inherent inconsistency, and here is a further one.

> "Major differences have been observed between fossil algae cells and the cells of modern algae. Why is that? At the same time, the 700-million-year-old fossils of soft-bodied animals like worms and jellyfish . . . resemble precisely the worms and jellyfish living on Earth today. Why haven't they changed with time?"[44]

The latter question is answered with a question: why should they change? The fact of non-change in some life forms is a puzzlement only so long as the entire fossil and genetic experimentation record is not heeded, only so long as the assumption of unlimited mutability/endless transformation continues to be held. There is no basis for retaining that assumption. Darwin himself would agree, I believe. In 1863, in a letter admitting that his evolutionary theory was only a general hypothesis, he said: "When we descend to details, we can prove that no one species has changed (i.e., we cannot prove that a single species has changed); nor can we prove that the supposed changes are beneficial, which is the ground work of the theory. Nor can we explain why some species have changed and others have not.

---

43. *Ibid.*, p. 30.

44. *Washington Post,* Oct. 3, 1977. The referenced worms and jellyfish are examples of "living fossils" or "persistent types" which have neither changed nor become extinct over hundreds of millions of years. They have been an enigma to evolutionists ever since Darwin's time. See content of note 45.

The latter case seems to me hardly more difficult to understand precisely and in detail than the former case of supposed change."[45]

As noted earlier, the discipline of paleontology is "in ferment". It is likewise for the sister subject of biology, especially as it pertains to man. The time is highly propitious for re-asking, in both schools, old questions (why do only some forms of life change?) as well as investigating new ones.

Among other things, the time is at hand to eliminate the gross discrepancies between concepts of evolutionary change and established laws of science. For example, and citing Wysong's work again (though Dewar and many others make the identical point), ". . . the second law [of thermodynamics] not only contradicts spontaneous origins, but also contradicts evolutionary transformations. Evolution of the species [in the traditional sense] means constant and steady increase in order and complexity — the amoeba to man idea. The second law says no such thing can occur spontaneously."[46]

Rather than inherent conflict, the relationship between life and entropy may well be positive. For example, all animal life on this planet depends on plant life, directly or indirectly. And what do plants do? Plants serve as energy binders. They trap and bind loose, free, random energy emitted from the Sun. If the Sun were not running down, like the clock spring, life as we know it could not exist. It is either a great anomaly, or a reflection of the limits of traditionalist insight, or both, but life appears to *require* the presence of entropy. Boulding notes: "It is indeed the throughput of energy in the earth that has made evolution possible."[47] Like him, let us get on with trying to solve the puzzle — without impudence.

SURVIVAL PROCESS

Why have some forms of life survived, while many others have appeared and then disappeared? Per the creed, blind natural forces caused life to emerge from inert matter; then the blind process of natural selection took over and dictated the terms of and conditions for survival.

The example of the dinosaurs suffices to discredit traditional theorizing. It is not until the Triassic era that the dinosaurs make

45. Charles Darwin, cited by Clark and Bales, *Why Scientists Accept Evolution, op. cit.,* p. 95.

46. Wysong, *Controversy, op. cit.,* p. 259.

47. Boulding, *Ecodynamics, op. cit.,* p. 47.

their appearance. According to Dewar (and other authorities) there is no evidence that the dinosaurs had ancestors. Nothing in the fossils links them with any precursors. "They have as an order so isolated a position, and are so widely separated from all other reptiles in structure, that they have long been a puzzle to Paleontologists."[48]

The appearance of the dinosaurs, a bizarre assembly of awesome and immense reptiles, constitutes one mystery. Then another: after having exercised dominion for over 100 million years, the dinosaurs suddenly became extinct. Just as there is no fossil evidence to account for their appearance, there is nothing in traditionalist theory to account adequately for their sudden demise. Rationales such as loss of food supply or living space due to the shifting of tectonic plates appear very lame, especially as an account for the loss of marine reptiles.

It seems to this layman that a great deal of humility is in order when it comes to claiming understanding the evolutionary process, particularly with regard to contentions as to what accounts for the appearance or disappearance of life forms. Adverse environmental circumstances and changes undoubtedly have played a role. But what other influences have been or are at work?

Norman Macbeth does an excellent job in *Darwin Retried* of showing that traditionalist answers regarding species survival, such as natural selection and survival of the fittest, are, upon close examination, noninformative tautologies. When evolutionists are pressed to explain, in detail, how and why some species populations multiply, while others stay stable, and still others die out, no useful answers are forthcoming. Natural selection is ostensibly affirmed through differential reproduction but the cause of differential reproduction remains unknown. Similarly, "fittest" is defined in terms of survival. "But this means that a species survives because it is the fittest and it is the fittest because it survives, which is circular reasoning and equivalent to saying that whatever is, is fit. The gist is that some survive and some die, but we knew this at the outset. Nothing has been explained."[49]

Bringing the occurrence of genetic mutations into the discussion adds nothing in the way of explanation regarding survival of life forms. Just the opposite. Wysong shows that when neo-Darwinism, which combines mutations and natural selection, is examined closely

48. Williston, cited in *Illusion, op. cit.,* p. 47.

49. Norman Macbeth, *Darwin Retried,* (Boston, Massachusetts: Gambit Incorporated, 1971), pp. 62-63.

it is exposed as a double negative.

"Natural selection is a weeding out of organisms, it is a destructive process. Further, what does nature have to select from? It is said mutations produce the new improved variants. Yet mutations are not known to increase the viability of organisms in natural environments. On the contrary, mutations, for all intents and purposes, are detrimental. Then, if natural selection cannot produce anything new, but only destroys organisms not fit (negative force), and mutations cannot produce new organisms, but only less fit ones (negative force), would not natural selection and mutations summated exert a negative force?"[50]

In debating the matter of what accounts for life survival or demise, I think it essential to appreciate C.S. Lewis' interpretation of miracles as recorded in the Bible.[51] I bring Lewis' work into play fully conscious of the reaction of derision it could draw, especially from ardent devotees of materialistic scientism. They are the most adamant defenders of "naturalistic" explanations and interpretations, even when it requires adhering to contradictory, wildly improbable or tautologic "answers". For them, *any* naturalistic interpretation, regardless how strained, gets preference. It demonstrates an "all or nothing" attitude, akin to the Darwinian belief that either every trivial detail of life was designed or nothing was designed. We must beware of getting trapped in such either/or fallacies.

I do not expect a frivolous reaction to Lewis' work by serious persons (scientists or others) concerned for the advancement of human understanding. Lewis shows that biblically-recorded miracles have a very definite pattern. They are not capricious happenings, merely arbitrary interferences scattered about. *They always pertain to implementation of a strategy.*

Evolution is the type of issue which cannot be successfully dealt with on a purely scientific basis. It requires a broad conceptual framework to lend meaning to the myriad of individual facts-of-the-matter. It is a preeminent example of what Mortimer J. Adler has acutely defined as a "mixed question," one requiring a blend of knowledges (scientific, historical, philosophical and, in this particular case, theologic) if true understanding is to result from

---

50. Wysong, *Controversy, op. cit.,* pp. 319-20.

51. C. S. Lewis, *Miracles: A Preliminary Study* (New York: MacMillan Publishers, 1974).

analytical effort.[52] Given that, it is perfectly in order to incorporate a work such as Lewis' in our thinking. Further, I suggest that portions of the paleontological record are plainly inexplicable absent considerations of Creator "strategy". I speculate that at least a portion of the fossil record is available to demonstrate to man what the Creator was *not* interested in finally sustaining: dinosaurs — great-sized, immensely powerful but *dumb* beasts. I am strongly inclined to think their remains give a fossilous clue vital to man's prospects in this militant, nuclear epoch.

How and why the survival process has operated remains shrouded in deep mystery. But there is substantial evidence that the process does not operate on the sole (traditionalist) principle of vicious, predatory domination. Dewar, Macbeth, and Wysong all illustrate evidence to the contrary; and a good portion of Leakey's book is devoted to demonstrating that "war" is not implanted in animal genes, nor human.

Of course instances of predatory behavior can be cited in the animal kingdom, as well as in the human. But barbarism and civilization have, and do, exist side-by-side. They are not mutually exclusive stages. In addition, I believe Leakey is right in his portrayal of cooperation in the evolution of society taking heavy precedence over crudely domineering behavior.

> "Unquestionably we are part of the animal kingdom. And, yes, at some point in our evolution we departed from the common dietary habits of the large primates and took to including a significant amount of meat in our menu. But a serious biological interpretation of these facts does *not* lead to the conclusion that, because once the whole of the human race indulged in hunting as part of its way of life, killing is in our genes. Indeed, we argue that the opposite is true, that humans could not have evolved in the remarkable way in which we undoubtedly have unless our ancestors were strongly cooperative creatures. The key to the transformation of a social apelike creature into a cultural animal living in a highly structured and organized society is sharing: the sharing of jobs and the sharing of food. Meat eating was important in propelling our ancestors along the road to humanity, but only as part of a package of socially-oriented changes involving the gather-

---

52. Mortimer J. Adler, *The Conditions of Philosophy: Its Checkered Past, Its Present Disorder, and Its Future Promise* (New York: Atheneum, 1965).

ing of plant foods and sharing the spoils.

This being so, why then is recent human history characterized by conflict rather than compassion? We suggest that the answer to this question lies in the change in ways of life from hunting to gathering to farming, a change which began about ten thousand years ago and which involved a dramatic alteration in the relationship people had both with the world around them and between each other. The hunter-gatherer is a part of the natural order; a farmer necessarily distorts that order. But more important, sedentary farming communities have the opportunity to accumulate possessions, and having done so they must protect them. This is the key to human conflict, and it is greatly exaggerated in the highly materialistic world in which we now live."[53]

Boulding likewise rejects traditionalist arguments. "The image of the biological realm as a kind of Hobbesian nightmare full of snarling species fighting to the death is a gross misunderstanding . . . Darwin's unfortunate metaphor of the 'struggle for existence' is a very poor description of the immense complexity of ecological interaction and the enormous numbers of strategies for survival in an evolutionary process . . . The struggle for existence . . . [is] a completely misleading metaphor. These processes are rare in the biological world . . . What might be called the 'original sin' school of sociobiologists — Lorenz, Ardrey, Tiger and Fox — placed a good deal of stress on the genetic origins of aggressiveness, fighting, and the use of threats in the human race. On the whole this seems very ill founded. Genetically, we are probably just as much related to the gentle democratic gibbon as to the baboon with its snarling machismo."[54]

Obviously, factors in addition to artifact protection, not the least of them paradigm conflict, must be taken into account in this era of ceaseless violence and turmoil. But that is not the point of note here. That which is stressed is that there exists no genetic, "programmed excuse" if nuclear-armed humans foil the established survival process.

HOMO SAPIENS: ANYTHING SPECIAL?

Linnaeus, who originated the modern system of biological

53. Leakey and Lewin, *Origins, op. cit.,* pp. 10-11 (emphasis in original).
54. Boulding, *Ecodynamics, op. cit.,* pp. 255, 21, 140, 154-5.

classification in the eighteenth century, could determine no single category appropriate for man and judged him both *common* and *unique*. I agree, and urge balanced attention be paid to both aspects. The traditional evolutionists, though, stress only the former characteristics of man. They argue that man is essentially nothing but a common "naked ape".

Their argument is supported, initially, by undeniable resemblances between man and the primates. But a look beneath the surface confirms that physical resemblances can be misleading, and that they provide no support for the traditionalist contention that man is merely a descendant of the apes. Dewar explains the distribution of anatomical characters among members of orders and higher groups. If all members of an order or other groups be derived from a common ancestor, it should be quite feasible to draw up a genealogical table showing the descent of each species from the common ancestor. For example, a common ancestor gives off descendants which evolve in different directions, one having a hairy integument, another having a wooly body-covering. These, in turn, give off lines of descent which continue to evolve separately. The anatomical characters of the hairy line may be designated a-b-c; those of the wooly line x-y-z. In no individual should both a-b-c and x-y-z characters appear.

However, Dewar points out, in no known group are anatomical characters distributed in such fashion. They are substantially mixed.

Man, for example, is said to have evolved from an old world monkey, in contrast to new world monkeys. Yet man exhibits a small percentage of characters similar to *both* old and new world monkeys. Taking a number of selected points, it can be shown that man shares a limited number of specific characteristics with a range of animals. Dewar notes that:

> "All this is not in accord with [traditionalist] theory . . . characters are curiously scattered.
>
> The primates are not peculiar in this respect. Every other group of animals of which the comparative anatomy has been studied exhibits similar features."[55]

Forms of life are so multiform that convergent resemblances could occur, evolution or no. Different species may show physical resemblances even when they are not genetically related, that is, not part of the same family. In some respects, apes are man's closest relation; in other respects not. Dewar reports that blood precipitation

---

55. Dewar, *Illusion, op. cit.,* pp. 246-7.

tests show that some *whales* are more nearly related to man than some monkeys are.

Nevertheless, evidence contrary to assumptions, as just mentioned, caused no amendment to traditionalist straight-line-of-descent theory during Dewar's time. He cites many instances of the kind of behavior warned against by Leakey (above): fudging of evidence and twisting of fact so that they might conform to predetermined "answers". Long ago, what began as the science of evolution degenerated, as far as the traditionalists are concerned, into activity aimed at rationalizing support for all of Darwin's original assumptions. It presents a very bad intellectual scene. It is, as Dewar put it, "an example of the way a Science suffers when it adopts a creed".[56]

The search for "origins" continues, and without objection so far as I am concerned. I just think it is totally futile, for reasons already given and others to follow, to develop an intellectually sound and satisfying understanding of how *Homo sapiens* came to be, if concepts and theory restrict attention to only natural phenomena independently transpiring.

Let us turn next to the archeological/paleontological findings of the Leakey family and their interpretation of certain fossils that have a bearing on this essay. We shall also consider the interpretive work of paleontologist Stephen Jay Gould in an effort to understand the evolution of the human family and to discern its original roots.

The Leakey family (father, mother, son) have devoted decades to ferreting evidence out of fossil volcanic ash in Africa. They have generated a conceptual revolution of evolution as it pertains to man. In the first place, they have rendered obsolete the "ladder" theory of evolution, defined by Gould as the popular picture of evolution as a continuous sequence of ancestors and descendants. The Leakey fossil finds indicate that many kinds of hominids (potential physical precursors) coexisted and that the prospective "shell" of man was formed a very long time back — now estimated at 3-4 million years ago. Gould writes:

> ". . . in 1964, Louis Leakey and his colleagues began a radical reassessment of human evolution by naming a new species from East Africa, *Homo habilis*. They believed that *H. habilis* was a contemporary of two australopithecine lineages; moreover, as the name implies, they regarded it as distinctly more human than either of its

---

56. *Ibid.*, p. 86.

contemporaries. Bad news for the ladder: three coexisting lineages of prehumans! And a potential descendent *(H. habilis)* living at the same time as its presumed ancestors. Leakey proclaimed the obvious heresy: both lineages of australopithecines are side branches with no direct role in the evolution of *Homo sapiens.*''[57]

Gould's work, *Ever Since Darwin,* is a collection of revealing essays by a committed, contemporary naturalist. The work has been well-received by reviewers and is advertised as ''a remarkable achievement''. The collection of essays is useful reading for anyone seriously interested in paradigm overhaul, notwithstanding his groundless aspersions of religion and human soul. One of the essays, ''Bushes and Ladders in Human Evolution,'' is particularly germane to the problem of discerning man's origin.

''At this point, I confess, I cringe, knowing full well what all the creationists who deluge me with letters must be thinking. 'So Gould admits that we can trace no evolutionary ladder among early African hominids; species appear and later disappear, looking no different from their great-grandfathers. Sounds like special creation to me.' . . . I suggest that the fault is not with evolution itself, but with a false picture of its operation that most of us hold — namely, the ladder; which brings me to the subject of bushes.

I want to argue that the 'sudden' appearance of species in the fossil record and our failure to note subsequent evolutionary change within them is the proper prediction of evolutionary theory as we understand it. Evolution usually proceeds by 'speciation' — the splitting of one lineage from parental stock — not by the slow and steady transformation of these large parental stocks. Repeated episodes of speciation produce a bush. Evolutionary 'sequences' are not rungs on a ladder, but a retrospective reconstruction of a circuitous path running like a labyrinth, branch to branch, from the base of the bush to the lineage now surviving at its top . . .

Homo sapiens is not the foreordained product of a ladder that was reaching toward our exalted estate from the start. We are merely the surviving branch of a once lux-

57. Stephen Jay Gould, *Ever Since Darwin: Reflections In Natural History* (New York: W. W. Norton and Co., 1977), p. 59.

urious bush.''[58]

If we accept the concept of speciation, vast puzzlements are resolved. Dewar's report on the "message of the rocks" would be seen as an accurate report on the workings of strictly natural phenomena, rather than the cause of wonderment about the intermittent involvement of a Creator.

Gould believes that to be the case exactly. He is a developer of the concept of "punctuated equilibria".[59] The content of the concept takes Darwin hard to task for his literally blind devotion to uniformitarianism. Gould agrees with those who, like Thomas Huxley, the first prominent defender of evolution, state that the paleontological record puts — and put even in Darwin's day — the lie to contentions of life never unduly disturbed, of slow, steady transformation as the unexceptionable rule. The concept of "punctuated equilibria" is set forth as the means of reconciling the physical record and a naturalist-only explanation of the evolution of life forms. Abrupt replacement of the forms of life is, per Gould, now to be finally appreciated as the evolutionary norm.

Directly related are Gould's offerings regarding speciation. They would explain the otherwise curious fact of mixed anatomical characteristics.

". . . major genetic reorganizations almost always take place in the small, peripherally isolated populations that form new species.

If evolution almost always occurs by rapid speciation in small, peripheral isolates — rather than by slow change in large central populations — then what should the fossil record look like? . . . We will first meet the new species as a fossil when it reinvades the ancestral range and becomes a large central population in its own right. [Genetic reorganization will preclude intermediate/transitional fossils.] . . . the wings of birds and insects, for example; the common ancestor of both groups lacked wings . . . During its recorded history in the fossil record we should expect no major change; for we know it as only a successful, central population.''[60]

58. *Ibid.,* pp. 60-62.
59. S. J. Gould and N. Eldredge, "Punctuated Equilibria: The Tempo and Mode of Evolution Reconsidered," *Paleobiology,* vol. 3, no. 2, Spring, 1977. Also, Gould in *The New York Times,* January 22, 1978, "Evolution: Explosion, Not Ascent."
60. Gould, *Ever Since Darwin, op. cit.,* pp. 61-62, 245.

One objective of the traditional evolutionists was to relieve man of "burdens of irrationality" regarding his origin and to dispense with supposedly primitive beliefs, as in miracles or the need to exercise religious faith. As noted specifically in *Illusion,* they failed in their efforts. At most, they only succeeded in substituting miracles of transformation for miracles of creation. The reader must judge if the contemporary Gould and his advocacy of speciation, which is a companion to his theory of punctuated equilibria, accounts convincingly for all basic occurrences and of changes in life forms. Has he improved on the record of the traditionalists?

If speciation were an observable transaction, it could well put an end to lingering dispute. But, it is crucially important to recognize that it is not.

"We are not likely to detect the event of speciation itself. It happens too fast, in too small a group, isolated too far from the ancestral range."[61]

What he is contending is that, — on evolutionary occasion, in an *unviewable* "out there" — elements of life engage, entirely on their own volition in *wholesale* genetic reorganization. Then presto! — unprecedented forms of life come on line as new central populations.

Maybe speciation works as Gould hypothesizes. I don't believe so, if for no other reason than that the DNA information code could not be sustained under conditions of "natural" massive scrambling. Further, if you believe as Gould does, admit candidly that you are not engaged in scientific exposition. You are only demonstrating the employment of "faith" — and on a massive scale. I submit that there is not a single element of Christianity which demands as much faith for acceptance as is required for the acceptance of speciation and its influence, as defined by Gould.

The questions are: when, and how, was the unique *habilis* "transformed" into the even more unique *sapiens?*

No one now knows, and we may never learn, a precise answer as to when man — as man — came to be. Creationists, in particular, must stop pretending otherwise. There is no sensible hope of deriving "the date" from genealogical exercises predicated upon the Book of *Genesis.*

The answer which might be given as to how *sapiens* came to be will depend, entirely, on which "faith" the reader chooses to employ:

---

61. *Ibid.,* p. 62.

that of the naturalist or that of the creationist.[62] And a decision is required, for *sapiens* is in part common but in most important part unique. Gould joins in the universal recognition that the basis for the operation of man's distinct characteristics is the fact that *sapiens* is disproportionately brainy (as distinguished from being disproportionately wise as a species — at least so far).

"The culprit in this tale is our most important evolutionary specialization, our large brain . . . My conclusion is not unconventional, and it does reinforce an ego we would do well to deflate. Nonetheless, our brain has undergone a true increase in size, not related to the demands of our larger body. We are, indeed, smarter than we were . . . But why did such a large brain evolve . . . And with this provocative query, I end, for we simply do not know the answer to one of the most important questions we can ask."[63]

Repeating certain points from Dewar: the fossil record shows that the earliest fossils of each class and order are not half-made or half-developed forms, but exhibit, fully developed, the characteristics of their class or order. Any changes which have been recorded have been very gradual and strictly limited, with no transgression of the limits of natural family. The evidence compiled by Wysong further supports this point.

Sapiens (man) has been no exception to those rules. Man came on the scene through a long-prepared "shell", but different and

---

62. Gould is not the only naturalist to advocate concepts requiring "faith" for their acceptance. It's been required from the very beginning of Darwinian evolutionary hypothesizing. Wysong, *Controversy, op. cit.,* at p. 419 gives a summary of "faith points" in addition to those of "punctuated equilibria" and "speciation."
  "Evolution requires plenty of faith: a faith in L-proteins that defy chance formation: a faith in the formation of DNA codes which if generated spontaneously would spell only pandemonium; a faith in a primitive environment that in reality would fiendishly devour any chemical precursors to life; a faith in experiments that prove nothing but the need for intelligence in the beginning; a faith in a primitive ocean that would not thicken but would only hopelessly dilute chemicals; a faith in natural laws including the laws of thermodynamics and biogenesis that actually deny the possibility for the spontaneous generation of life; a faith in future scientific revelations that when realized always seem to present more dilemmas to the evolutionist; faith in probabilities that treasonously tell two stories — one denying evolution, the other confirming the creator; faith in transformations that remain fixed; faith in mutations and natural selection that add to a double negative for evolution; faith in fossils that embarrassingly show fixity through time, regular absence of transitional forms and striking testimony to a world-wide water deluge; a faith in time which proves to only promote degradation in the absence of mind; and faith in reductionism that ends up reducing the materialist's arguments to zero and forcing the need to invoke a supernatural creator."
63. Gould, *Ever Since Darwin, op. cit.,* pp. 75, 184-5, 191.

"whole", and has undergone no change of consequence since his initial appearance. No one has better illuminated the point than G.K. Chesterton through his classic work *The Everlasting Man*.

In an early chapter, examining the meaning of what remains from "The Man in the Cave," Chesterton depicts central lessons to be appreciated, lessons so simple that the traditional evolutionists have completely overlooked them, intentionally or otherwise.

". . . The cave might have had a special purpose like the cellar; it might have been a religious shrine or a refuge in war or the meeting-place of a secret society or all sorts of things. But it is quite true that its artistic decoration has much more of the atmosphere of a nursery than of any of these nightmares of anarchial fury and fear. I have conceived a child as standing in the cave; and it is easy to conceive any child, modern or immeasurably remote, as making a living gesture as if to pat the painted beasts upon the wall. In that gesture there is a foreshadowing, as we shall see later, of another cavern and another child.

But suppose the boy had not been taught by a priest but by a professor, by one of the professors who simplify the relationship of men and beasts to a mere evolutionary variation. Suppose the boy saw himself, with the same simplicity and sincerity, as a mere Mowgli running with the pack of nature and roughly indistinguishable from the rest save by a relative and recent variation. What would be for him the simplest lesson of that strange stone picture-book? After all, it would come back to this; that he had dug very deep and found the place where a man had drawn the picture of a reindeer. But he would dig a good deal deeper before he found a place where a reindeer had drawn a picture of a man. That sounds like a truism, but in this connection it is really a very tremendous truth. He might descend to depths unthinkable, he might sink into sunken continents as strange as remote starts, he might find himself in the inside of the world as far from men as the other side of the moon; he might see in those cold chasms or colossal terraces of stone, traced in the faint hieroglyphic of the fossil, the ruins of lost dynasties of biological life, rather like the ruins of successive creations and separate universes than the stages in the story of one.

He would find the train of monsters blindly developing in directions outside all our common imagery of fish and bird; groping and grasping and touching life with every extravagant elongation of horn and tongue and tentacle; growing a forest of fantastic caricatures of the claw and the fin and the finger. But nowhere would he find one finger that had traced one significant line upon the sand; nowhere one claw that had even begun to scratch the faint suggestion of a form. To all appearances, the thing would be as unthinkable in all those countless cosmic variations of forgotten aeons as it would be in the beasts and birds before our eyes. The child would no more expect to see it than to see the cat scratch on the wall a vindictive caricature of the dog. The childish common sense would keep the most evolutionary child from expecting to see anything like that; yet in the traces of the rude and recently evolved ancestors of humanity he would have seen exactly that. It must surely strike him as strange that men so remote from him should be so near, and that beasts so near to him should be so remote. To his simplicity it must seem at least odd that he could not find any trace of the beginning of any arts among any animals. That is the simplest lesson to learn in the cavern of the coloured pictures; only it is too simple to be learnt. It is the simple truth that man does differ from the brutes in kind and not in degree; and the proof of it is here; that it sounds like a truism to say that the most primitive man drew a picture of a monkey and that it sounds like a joke to say that the most intelligent monkey drew a picture of a man. Something of division and disproportion has appeared; and it is unique. Art is the signature of man . . .

It is not natural to see man as a natural product. It is not common sense to call man a common object of the country or the seashore. It is not seeing straight to see him as an animal . . . It sins against the light; against that broad daylight of proportion which is the principle of all reality. It is reached by stretching a point, by making out a case, by artificially selecting a certain light and shade, by bringing into prominence the lesser or lower things which may happen to be similar. The solid thing standing in the sunlight, the thing we can walk around and see from all sides, is quite different. It is also quite extraordinary; and the more sides we see of it the more extraordinary it seems. It is emphatically not a thing that follows or flows naturally from anything else."[64]

Repeating, we do not presently know, and may never learn the precise date when *Homo sapiens* first/finally came into being. But paleoanthropologist David Pilbeam, citing recently found fossil evidence from Pakistan and India, affirms Gould's concept that ". . . hominoid evolution . . . no longer resembles a ladder but is, instead, more like a bush. The story is diverse and subtle and not at all straightforward. Early species are not simply pale copies of later ones, predestined to evolve into their descendents . . . (E)xtinct hominoids were not particularly modern. They were not like either living apes or human beings but instead were unique, distinct animal species . . . There is no clear-cut and inexorable pathway from ape to human being."[65] I suggest evolutionary "bush" and preparatory "shell" are vitally essential concepts for appreciating the eventual and sustained presence of the most unique species — man — whose first hallmark is an exceptional brain that supports completely unique abilities for propositional speech and conceptual thought,[66] whose signature is art, and to whose honored position in the order of things we shall return.

---

64. G. K. Chesterton, *The Everlasting Man,* (New York: Dodd, Mead and Company, Apollo Edition, 1971), pp. 14-15, 20.

65. Pilbeam, "Rearranging Our Family Tree," *op. cit.,* pp. 44, 42.

66. These central distinctions are explained in masterful fashion by Adler in *The Difference Of Man And The Difference It Makes, op. cit.* Also of great importance to the subject at hand is Adler's *What Man Has Made Of Man* (New York: Frederick Ungar Publishers, 1957), especially his exposure of the ways Freudian psychological concepts are reflective (in important part) of wildly erroneous evolutionary theorizing.

It is also useful, at this point, to heed the cautioning of S.L. Washburn regarding the limitations of sociobiology in its present state, "What We Can't Learn About People From Apes," *Human Nature,* November, 1978, p. 75.

". . . Language gives the nature versus nurture [or genetic versus learning] controversy over human behavior an entirely new dimension. No comparable mechanism, nor any alternative biological system, allows virtually unlimited communication as well as the development of new symbols when needed . . . In applying evolutionary theory to human conduct, sociobiologists generally make little effort to understand the complexity of human behavior. Instead they use the theory of evolution to make questionable comparisons between human and animal behavior without the necessary research to validate their claims. Unsubstantiated opinions by researchers are too often presented as facts — or at least worthy contributions to scholarship. I would be the first to agree that the full understanding of the behavior pattern of any species must include biology. But the more learning that is involved, the less there will be of any simple relation between basic biology and behavior. The laws of genetics are not the laws of learning. As a result of intelligence and speech, human beings provide the extreme example of highly varied behavior that is learned and executed by the same fundamental biology. Biology determines the basic need for food, but not the innumerable ways in which this need may be met."

PURPOSELESS UNIVERSE/LIFE?

Why is there anything at all? Why not nothing? What is the meaning, the purpose of this universe, of human life?

These are among the greatest metaphysical questions of all time; and the traditionalist answer to them is that this universe just happens to exist, with no particular meaning or purpose attached to its presence. Likewise for human life — there is no manifest point to it. Thus contends the traditionalist. This philosophical position results from an extension of Darwin's insistence that the process of natural selection in nature operates without discernible purpose (except possibly for the purpose of mere survival fitness).

The traditionalist answers are not very satisfying. Steven Weinberg closes *The First Three Minutes* with a common lament: "The more the universe seems comprehensible, the more it also seems pointless . . . The effort to understand the universe is one of the very few things that lifts human life a little above the level of farce, and gives it some of the grace of tragedy."[67]

I believe that we can now put an end to such lamenting, and do so with more assurance and more comprehensively than at any previous time in history.

I take it as now beyond dispute, scientifically or philosophically, that the universe had a start, that there is something rather than nothing because a Creator wants it to be so.

That Creator exhibits through His works a supreme intelligence. Repeating the characterization of C. S. Lewis: "What is behind the universe is more like a mind than it is like anything else that we know."

While urging that a Biblical religious orientation be reflected in an overhauled world/life paradigm, I readily concede that the demonstrable presence of Intelligent Creator does not immediately equate with the God of the Bible. Nor does the presence of Intelligent Creator automatically illuminate Purpose, either of the Creation or of human life. But I do submit that the necessary "bridges" between

---

Washburn succinctly explains the essential differences between human and monkey "communication," and, opposite of Gould, highlights the limitations of the monkeys/apes in this regard. And I think it worth noting that both Adler *(Difference)* and Boulding *(Ecodynamics)* convey that the animal apparently closest to *Homo sapiens* in brain-body ratio and communication ability is the *bottlenosed dolphin.* (Washburn's points are explained in fuller detail in *Human Evolution: Biosocial Perspectives* (Perspectives on Human Evolution, Vol. IV; Benjamin/Cummings Publishing Co., Menlo Park, Calif., 1978), edited by S. L. Washburn and Elizabeth R. McCown.)

67. Weinberg, *First Three Minutes, op. cit.,* pp. 154-55.

these points have already been built, are in physical and intellectual place, and are as immediately and fully operational as man chooses to make them.

Is the Creator the God of the Bible? He can be *no one else.* That conclusion is supported by two types of complementary evidence, either of which I believe sufficient.

—First, the great religions of the world are not in any conflict over the matter. The Eastern religions simply do not deal, in a formalistic way, with the issue of the origin of the universe or of human life. They comment upon, and teach various suppositions about, eternity. Yet, they are of no help in trying to discern what is behind, say, the residual microwave radiation.

There is no conflict between the remaining religions — the monotheistic religions of Islam, Judaism, and Christianity — for the simplest of reasons. The starting point for all three is the Book of *Genesis.* They are obviously in agreement as to Who was responsible for the Creation.

—Second, Etienne Gilson's classic work, *God and Philosophy,* contains a masterfully penetrating historical examination of the question at hand and arrives at the same answer.[68] Following is a summary of that work.

Gilson begins by examining the religious and philosophical state of affairs in ancient Greece. Plato and Aristotle led the Greeks out of the intellectual woods of myriad, personally threatening deities and "forces". But there they ran into the limit, for the first time, of human reason. An ultimate, impersonal First Cause came to be postulated but no contribution beyond that was made toward comprehension of universal origins.

The Mediterranean region was rife with religio-mythical competition in the era of the great Greek thinkers. But there was a large sameness about them all — save for one, that of the ancient Jews. Gilson depicts the startling difference. The religion of the Jews was predicated not upon myth-symbolism (though they often lapsed into such), nor upon refined conceptual achievements such as First Cause. Judaism was based upon the belief in a Creator, one committed to giving special attention to the Jewish tribes. He was no idle abstract. A live, present Being is the God of the Jews: "I am Whom am." Moses received and spread the direct word of Yahweh, which name means "He who is".

---

68. Etienne Gilson, *God and Philosophy* (Clinton, Massachusetts: Colonial Press, Inc., Nineteenth Printing, 1976).

The Jews had not pondered "origins," yet they claimed to know what had transpired. The Greeks had pondered, had arrived at the object of First Cause, but could say nothing more about how or why the universe came to pass. Is it possible to reconcile the Logical and the Revealed? Gilson points out the dividing line between Greek and Judeo-Christian thought. Moses said: *He who is;* Plato said: *That which is.* The difference introduces an entirely new philosophical problem. The Greek asks: What is nature? The Christian asks: What is being?[69]

Gilson shows that it took centuries of labor to overcome the faulty start of Augustine and achieve a viable reconciliation of reason and revelation. It was not until Thomas Aquinas that the hardest questions of all — what is "being"?; what do you mean by a God who "is"? — finally got a definitive answer.

> Why had the Greek mind spontaneously stopped at the notion of nature, or of essence, as at an ultimate human experience? Because, in our human experience, existence is always that of a particular essence . . . [But] the essence of any and every thing is not existence itself, but only one of the many possible sharings in the existence. This fact is best expressed by the fundamental distinction of "being" and "what is" so clearly laid down by Thomas Aquinas. It does not mean that existence is distinct from essence as a thing from another thing . . . Existence is not a thing, but the act that causes a thing both to be and to be what it is . . .
>
> [What the natural] sciences cannot teach us is why this world, taken together with its laws, its order, its intelligibility, is, or exists. If the nature of no known thing is "to be," the nature of no known thing contains in itself the sufficient reason for its own existence. But it points to its sole conceivable cause. Beyond a world wherein "to be" is everywhere at hand, and where every nature can account for what other natures are but not for their common existence, there must be some cause whose very essence it is "to be." To posit such a being whose essence is a pure Act of existing, that is, whose essence is not to be this and that, but "to be," is also to posit the Christian God as the supreme cause of the universe.[70]

69. *Ibid.,* pp. 42, 44.
70. *Ibid.,* pp. 70, 71, 72.

It is a terrible irony that, virtually as soon as Aquinas achieved the reconciliation between reason and revelation, the accomplishment came to be disregarded. Gilson explains how the magnificent feat fell victim to the philosophical revolution instigated by Descartes, a revolution based on the premise that theology and metaphysics lack intrinsic value for the world of practical affairs, that the only useful methods for securing reliable knowledge are those which have come to be developed in conjunction with the natural sciences. Without impugning in any fashion the legitimate techniques of scientific investigation, it is still necessary to state that the dismissal of metaphysics and theology as irrelevant has proven to be about the biggest mistake in man's mistake-filled history. Why the dismissal constitutes an error of legendary proportions will be shown shortly. Here the task is to conclude summation of Gilson's work.

Since the time of Descartes and on, the reconciliation of Aquinas has been widely ignored. But no one has ever successfully challenged it.

> [For] the God of natural theology, true metaphysics does not culminate in a concept, be it that of Thought, of Good, of One, or of Substance. It does not even culminate in an essence, be it that of Being itself. Its last word is not *ens,* but *esse;* not *being,* but *is.* The ultimate effort of true metaphysics is to posit an act by an Act, that is, to posit by an act of judging the supreme act of existing whose very essence, because it is to be, passes human understanding. Where a man's metaphysics comes to an end, his religion begins. But the only path which can lead him to the point where the true religion begins must of necessity lead him beyond the contemplation of essences, up to the very mystery of existence. This path is not very hard to find, but few are those who dare to follow it to the end. Seduced as they are by the intelligible beauty of science, many men lose all taste for metaphysics and religion. A few others, absorbed in the contemplation of some supreme cause, become aware that metaphysics and religion should ultimately meet, but they cannot tell how or where; hence they separate religion from philosophy, or else they renounce religion for philosophy, if they do not, like Pascal, renounce philosophy for religion. Why should we not keep truth, and keep it whole? It can be done. But only those can do it who realize that He Who is the God of the

philosophers is HE WHO IS, the God of Abraham, of Isaac, and of Jacob.[71]

Chesterton's *Everlasting Man* is also relevant to this most central matter. The following quotation portrays vividly the distinguishing origins of Christianity.

". . . Divine Plato, like Divus Caesar, was a title and not a dogma. In Asia, where the atmosphere was more mythological, the man was made to look more like a myth, but he remained a man. He remained a man of a certain special class or school of men, receiving and deserving great honour from mankind. It is the order or school of the philosophers; the men who have set themselves seriously to trace the order across any apparent chaos in the vision of life. Instead of living on imaginative rumours and remote traditions and the tail-end of exceptional experiences about the mind and meaning behind the world, they have tried in a sense to project the primary purpose of that mind *a priori*. They have tried to put on paper a possible plan of the world; almost as if the world were not yet made.

Right in the middle of all those things stands up an enormous exception. It is quite unlike anything else. It is a thing final like the trump of doom, though it is also a piece of good news; or news that seems too good to be true. It is nothing less than the loud assertion that this mysterious maker of the world has visited his world in person. It declares that really and even recently, or right in the middle of historic times there did walk into the world this original invisible being about whom the thinkers make theories and the mythologists hand down myths; the Man Who Made the World. That such a higher personality exists behind all things had indeed always been implied by all the best thinkers as well as by all the most beautiful legends. But nothing of this sort had ever been implied in any of them. It is simply false to say that the other sages and heroes had claimed to be that mysterious master and maker, of whom the world had dreamed and disputed. Not one of them had ever claimed to be anything of the sort. Not one of their sects or schools had ever claimed that they had claimed to be anything of the sort. The most that any religious proph-

---

71. *Ibid.*, pp. 143-44.

et had said was that he was the true servant of such a being. The most that any visionary had ever said was that men might catch glimpses of the glory of that spiritual being; or much more often of lesser spiritual beings. The most that any primitive myth had ever suggested was that the Creator was present at the Creation. But that the Creator was present at scenes a little subsequent to the supper-parties of Horace, and talked with tax collectors and government officials in the detailed daily life of the Roman Empire, and that this fact continued to be firmly asserted by the whole of that great civilization for more than a thousand years — that is something utterly unlike anything else in nature. It is the one great startling statement that man has made since he spoke his first articulate word, instead of barking like a dog. Its unique character can be used as an argument against it as well as for it. It would be easy to concentrate on it as a case of isolated insanity; but it makes nothing but dust and nonsense of comparative religion."[72]

Even if the God of the Bible is firmly established as Creator of the universe of which we are a part, His purpose for it and us is very obscure to very many. To clarify the Purpose, and make it comprehensible to all, is the objective of Chapter V of *The Direct Connection*. I summarize the Purpose here, and urge the reader to examine my arguments leading to the deceptively simple conclusion: the purpose of man spending tension-laced time on Earth is to give him the background experience necessary to appreciate the wonders of eternal Heaven (assuming he does as is required to earn the opportunity of Heaven). Man undergoes a *necessary training experience* in the earthly phase of life, and it *is* but the first phase of life as far as obedient man is concerned. And to preclude any unwarranted scoffing, it is worth adding a note of explicit assurance that "science" has not done and cannot do anything at all to add or detract from the potential of Heavenly eternity.

I want also to repeat that all major religions, all cultural regions of the globe can contribute to a properly revamped world/life paradigm. Each has something to contribute to moving the Creation "upward and outward" — in a social and spiritual sense, not merely an expanse of economic production. That kind of growth and maturation is feasible, notwithstanding the laws of entropy. Schumacher's *Guide* is an excellent beginning for understanding

---

72. Chesterton, *Everlasting Man, op, cit.,* pp. 334-5.

what is required for individual spiritual fulfillment and how contributions from the various religions can be effectively correlated towards that end.[73]

To close this section, a comment on the veracity of the Biblical message, and upon the relationship between the "first word" and the "last word" is in order.

The first word, the Book of *Genesis,* the foundation of creationist belief, was — and remains — the only alternative available to those who opt for the cause of Darwinian evolution.[74] The original proponents did not champion Darwinian concepts because they had been well-demonstrated. Not at all. Thomas Huxley, for example, "bulldoged" for Darwin because he judged the *Genesis* account, or that account as it had been and was being interpreted, palpably deficient in light of known evidence. Still, that same evidence, he admitted, did not confirm Darwinian evolution.

I believe Huxley's position can be fairly capsulized as follows. He was not an atheist, recognized that First Cause could not be ignored, and appreciated that human consciousness could not be explained through purely materialistic interpretations. For Huxley, God had to be involved in some way: "By the word God I understand a being absolutely infinite, that is, a constant [self-existing] substance with infinite attributes."

But Huxley could not abide the God of the Bible, at least as he understood Him. "I really believe that the alternative is either Darwinism or nothing, for I do not know of any rational conception or theory of the organic universe which has any scientific position at all beside Mr. Darwin's . . . Whatever may be the objections to his views, certainly all other theories are out of court . . . [The] fact of evolution is to my mind sufficiently evidenced by paleontology . . . [As for the lessons of *Genesis*] my sole point is to get the people who persist in regarding them as statements of fact to understand that they are fools . . . [But] I by no means suppose that the transmutation

---

73. Another important contribution in the same regard is *Food/Energy and the Major Faiths* (Report on the First Interreligious Peace Colloquium), Joseph Gremillion, editor (Maryknoll, N.Y.: Orbis Books, 1978).

74. Tom Bethell, concerned mainly with the political implications of changing views toward evolutionary theory, makes a good point in his previously-cited article ("Burning Darwin To Save Marx"): "If Darwin's theory were decisively underminded, it would still be possible to argue that evolution had taken place as a result of mechanisms not yet understood. Some scientists do take this position. Darwin debunked does not leave us with Genesis as the only alternative. [I disagree; see text.] Nevertheless, there are those who argue that the abandonment of the evolutionary mechanism would inevitably lead to doubts that evolution had occurred at all. *That* is undoubtedly why Darwin is still defended so stoutly — not because his supporters are capitalists but because they are materialists."

hypothesis is proven or anything like it. [I do] view it as a powerful instrument of research. Follow it out, and it will lead us somewhere; while the other notion is like all the modifications of 'final causation,' a barren virgin.''[75]

The problem of the misinterpretation of *Genesis* persists to this day. The most common misinterpretation, certainly not restricted to traditional evolutionists, is that time and advancement of scientific knowlege have rendered the *Genesis* account even more passé than judged by Huxley a century ago. As for fundamental creationists, to suggest to them that the *Genesis* account and an initial explosion are commonly related is to risk a highly contemptuous reaction from the other side.

Neither form of misinterpretation, however, should cause us to hesitate from making proper use of *Genesis.* To do that we must, first, keep fully in mind that the *Genesis* account of creation is allegory; that is, it portrays events only in symbolic or generally descriptive fashion. It does not convey statements of fact in normal scientific language. It gives an explanation, in lay language; one which man, through his intelligence, is able to test. That understood, let us compare the message of *Genesis* with the findings of honest science.

—Legitimate science tells us that the universe started with a big bang; that even after its formation the Earth was initially void of life; that what we know and experience as earthly life, in general and of human life specifically, is the result of sequential development.

—The message of the ancient book, *Genesis,* is, properly understood, the same as that of our best science. It agrees that there was a beginning.[76] It coincides precisely with the findings of the astrophysicists and paleontologists (discussed earlier) that the early Earth was without life. *Genesis* 1:2 is scientifically solid and as good a description of primeval conditions as can be produced. ". . . the earth was waste and void: and darkness moved upon the face of the deep: and the Spirit of God moved upon the face of the waters."

According to the Word received by Moses, the Earth was once

---

75. Thomas Huxley, cited by Clark and Bales, *Why Scientists Accept Evolution, op. cit.,* pp. 78, 81, 73.

76. Robert Jastrow has amplified upon the findings cited earlier. In his latest book, *God And The Astronomers* (New York: W. W. Norton, 1978), he states: "The details differ, but the essential elements in the astronomical and biblical accounts of Genesis are the same: the chain of events leading to man commenced suddenla and sharply at a definite moment in time, in a flash of light and energy." (p. 14).

lifeless, dark, uninhabitable. The Earth had to be prepared for life, which was brought into being through a sequence of Creator acts over an allegorically defined period of time, "six days".

James D. Bales has published a short but lucid and most worthwhile explanation of these matters, *The Genesis Account and A Scientific Test.*[77] Huxley's admonition to "follow it out" has been heeded. Since he cannot, will his disciples admit to the intellectual and behavioral consequences of traditionalist evolution having been "found out"?

Science affirms the "first word" of the Biblical, Christian message. That message is one which developed over historical time. Through the appearance of the prophets, and of Jesus, the message was presented — and modified — over time. All portions of the biblically recorded message cannot be granted equal weight, for some parts have been superseded. That portion of the message deserving the greatest weight is the "latest word" — as represented by the teaching of Jesus and the preaching of Paul — and emphatically not that represented by the Book of Revelation.[78]

The interpretation of evolution I have set out is as consistent with the "latest word" of God as it is with the "first word" and the actual findings of the natural sciences. It thereby fulfills the promise made near the outset of this essay: harmony is reachieved between the Word and works of God, erudite theology, legitimate philosophy[79] and honest science.

CONCLUSION

The philosophic revolution instigated by Descartes supported a great growth in compartmentalized scientific investigation and adherence to the method of reductionism. The purpose of science is to gain and apply knowledge. The stockpile of knowledge which has been accumulated over the last century is mainly attributable to the

---

77. James D. Bales, *The Genesis Account and a Scientific Test* (Searcy, Arkansas: 707 East Race, zip 72143, 1975).

78. As I explain in *Connection* (p. 102), if we hope to understand anything we *must* assume that order, not caprice, underlies our environment. "Just as we must assume that order governs the physical universe, it is correct to assume that God is also consistent in other revelations." The Book of Revelation is *not* consistent with the teachings of Jesus or the preaching of Paul. It is a patchwork job which comes out at large odds with the "latest word". For a scholarly dissection of Revelation see J. Massyngberde Ford, *Revelation: The Anchor Bible* (New York: Doubleday and Co., 1975).

79. By "legitimate philosophy" I refer to that type explained and defended by Adler in *The Conditions Of Philosophy, op. cit.* It will represent a new undertaking which can no longer be left largely unattended.

concentrated attention and narrow-focus methods of reductionism.

The knowledge which has resulted can in no way be denied or belittled. But we must admit that overapplication of reductionist methodology and outlook entails extremely severe problems. The great need of this historical hour is integration, synthesis, joining the pieces of reductionist expertise into a coherent whole under a rationally defensible philosophy. Just as unrelieved uniformitarianism will not suffice to explain evolution, neither will unaided reductionism serve to explain and guide human life.

Adherence to reductionist-only methods prevents our seeing the forest for the trees. The problem is particularly acute within the scientific community, and assuredly with regard to the evolutionary controversy. Returning again to Wysong:

> So why is evolution believed? Some believe because their scientific training and work is so highly specialized that they cannot through overview see the contradictory issues . . . Scientists in each narrowed corner simply assume that other scientists in other narrowed corners have the proofs all worked out. Few have ever seen in perspective the real issues . . . In fact, when renowned scientists are faced [with the contradictions in traditionalist concepts] they are usually dumbfounded and whirl in the disbelief that such information has not been circulated.[80]

It is long overdue to end the suppression, described by Wysong, of pro-creationist arguments. It is imperative that the issues of evolution be examined in their proper context. For in the name, and under the guise, of scientific objectivity, the traditional evolutionists, the developers and defenders of a totally materialistic evolutionary creed, have promoted methods of analysis and data interpretation which betray the very presence of human intelligence. They stand as king-sized examples of anti-intellectualism.

Having given close examination to the actual evidence as compared to traditionalist arguments, Dewar comments that "The credulity of transformists knows no limit!" and "[Their] mental gymnastics . . . have long been a source of wonder to me."[81] Even more damningly, Schumacher cites psychiatrist Karl Stern:

> If we present, for the sake of argument, the theory of evolution in a most scientific formulation, we have to say

---

80. Wysong, *Controversy, op. cit.,* p. 420.
81. Dewar, *Illusion, op. cit.,* pp. 216, 253.

something like this. "At a certain moment of time, the temperature of the Earth was such that it became most favourable for the aggregation of carbon atoms and oxygen with the nitrogen-hydrogen combination, and that from random occurrences of large clusters molecules occurred which were most favourably structured for the coming about of life, and from that point it went on through vast stretches of time, until through processes of natural selection a being finally occurred which is capable of choosing love over hate and justice over injustice, of writing poetry like that of Dante, composing music like that of Mozart, and making drawings like those of Leonardo." Of course, such a view of cosmogenesis is crazy. And I do not at all mean crazy in the sense of slangy invective but rather in the technical meaning of psychotic. Indeed such a view has much in common with certain aspects of schizophrenic thinking.[82]

But, Schumacher adds:

"The fact remains . . . that this kind of thinking continues to be offered as objective science not only to biologists but to everybody eager to find out the truth about the origin, meaning, and purpose of human existence on Earth, and that, in particular, all over the world virtually all children are subjected to indoctrination along these lines . . . [Traditional evolutionism] can be described as a peculiarly degraded religion, many of whose high priests do not even believe in what they proclaim . . . It has destroyed all faiths that pull mankind up and has substituted a faith that pulls mankind down."[83]

There are many intolerable aspects to the existing situation, one of them being textbooks and other influential publications which convey the impression of a united scientific front in support of traditionalist evolutionary concepts. That impression is false in large measure, and becomes more so each day. Macbeth, Dewar, Bales, Wysong, and Boulding all identify hosts of highly qualified critics. I cited authorities in opposition in *The Direct Connection.* Schumacher does likewise, pointing out that:

---

82. Karl Stern, cited by Schumacher in *Guide, op. cit.,* p. 113.
83. Schumacher, *Guide, op. cit.,* pp. 114-5.

Counterarguments are simply ignored. The article on "Evolution" in the *New Encyclopedia Britannica* (1975) concludes with a section entitled "The Acceptance of Evolution," which claims that "objections to evolution have come from theological and, for a time, from political standpoints." Who would suspect, reading this, that the most serious objections have been raised by numerous biologists and other scientists of unimpeachable credentials?[84]

One of the principal aims of this essay is to make a path, to create an opening for qualified, other-than-traditionalist spokesmen. There is a vast amount of rethinking and reconceptualizing to be done. We need to get in front of as large a public as possible, as fast as possible, learned, responsible viewpoints which can assist in firming up a sound alternative to the traditionalist paradigm.

Among other things, this means that the time has arrived for Immanuel Velikovsky to be given a *serious* hearing. The scientific community has brought much dishonour upon itself for past, and continuing, efforts to "blackball" and discredit out-of-hand Velikovsky's "non-conformist" concepts.[85] He raises fundamental questions about the Earth's evolutionary development which traditionalist theory cannot answer *properly,* and which need answering. I am not qualified to judge all of Velikovsky's theories, but I do side with him openly and completely on one crucial point: the physical evidence clearly indicates that the Earth's evolution involved times of massive, catastrophic change. The very presence of fossils refutes unrelieved uniformitarianism, for fossil formation generally necessitates rapid (catastrophic) burial. The evidence plainly indicates that major "discontinuities" have taken place.[86] This, in

---

84. *Ibid.,* p. 114.

85. See note 24. The critiques of Velikovsky's work presented at the 1974 AAAS session have been published by Cornell Press under the title *Scientists Confront Velikovsky.* The fraud continues in active operation. It has been further exposed in *Velikovsky and Establishment Science* (Glassboro, New Jersey: Kronos Press, 1977). An even-handed critique of Velikovsky's work is provided by C. J. Ransom in *The Age of Velikovsky* (Glassboro, New Jersey: Kronos Press, 1976). It is far superior to the (mis)interpretation of Velikovsky's basic effort set forth by Gould in *Ever Since Darwin, op. cit.* Also see KRONOS IV:2 (Winter-1978) — "Scientists Confront Scientists Who Confront Velikovsky" — which is a continuation of the special AAAS issue of KRONOS (III:2). It must be emphasized here, however, that Velikovsky himself is *not* a "fundamentalist" though his critics have often seen fit to label him as such. What Velikovsky proposes is the thesis of "natural revolution".

86. See Roger W. Wescott, "Polymathics and Catastrophism," KRONOS IV:1 (Fall - 1978), pp. 3-20.

itself, is sufficient justification for relegating traditionalist gradual change/uniformity theory to the doctrinal trashcan, "to the museum of exploded theories".

We must get on with paradigm overhaul in order, among other reasons, to better understand God-the-activist. We must not allow misinterpretation of *Genesis* to continue. God did not forever cease His work after "six days". The geologic record fully confirms Jesus' characterization of Him in this regard: "My Father has never yet ceased his work, and I am working too."[87] Our work must proceed deliberately, but with speed. We must insure that man-the-activist does not blunder into potentially gross violations of His Purpose.

---

87. John 5:17, *The New English Bible* (Oxford Univ. Press).

Chapter III

THE SECOND GENESIS:
THE COMING CONTROL OF LIFE

Chapter III

THE SECOND GENESIS:
THE COMING CONTROL OF LIFE

A primary objective of this book is to rid the educated world of the single most damaging of "accepted" teachings, that of the traditional evolutionists. The foregoing analysis should help set the stage for consideration of a comparable goal: to generate awareness of the indispensable role of erudite theology in combination with sound philosophical reasoning to guide the science-based and science-influenced portion of man's activities. In no realm is such guidance more urgently needed than that regarding potential manipulation of human life.

We are confronted by a new generation of transformists. Unlike those with whom Dewar had to deal — who were armed only with illusions — the new generation has at its disposal increasingly powerful technologies developed through the behavioral, biological, and computer sciences. Of most concern is the mind-set which seems dominant amongst the would-be transformers. It is heavily reflective of crude materialistic scientism. As Vance Packard, author of *The People Shapers,* observed:

> ". . . The same thread appears when we consider assumptions of pioneering activists from such diverse fields as reproductive biology, psychosurgery, and molecular biology. *What emerges is a pervading assumption that humans are creatures of almost limitless plasticity.* People are raw material that needs perfecting, modifying, or at least improving, either for their own good or to

suit the wishes of others . . . Whereas the old believers in the perfectability of people thought primarily in moral terms, the new [transformists] want to change people physically, emotionally, mentally . . ."[88]

The people shapers, if left to their own visions and devices, represent a supreme threat to humane life experience. As humanistic psychologist Carl Rogers acknowledges: "We can choose to use our growing knowledge to enslave people in ways never dreamed of before, depersonalizing them, controlling them by means so carefully selected that they will perhaps never be aware of their loss of personhood."[89]

In working against transformist threats, I believe the most immediate need is to establish a defensible context for evaluating potential applications of biomedicine. That begins with the title chosen for the present chapter, the same as that of a book by Albert Rosenfeld,[90] science editor of *Saturday Review*. He chose the title "in order to convey the radical nature of the biomedical revolution". I repeat the title for exactly the same reason. As Rosenfeld notes in a later article, "One thing biology is clearly telling us is that the human future will be quite different from the human [past and] present."[91]

We are witness to man stepping across an historic threshhold. The criticality of the move could not be more awesome.

Given what he terms "our spirited explorations along biology's frontiers," Rosenfeld notes that "we may have to take upon ourselves some godlike prerogatives as we become self-anointed trustees of our own evolution".[92]

The dimensions of the prospect are absolutely stunning. Even more: man rushing to take biological control of his own — and the planet's — evolution takes on a frightening proportion when it is realized, as it must be broadly realized

—that we do not yet have an adequate grasp on how evolution has actually proceeded, and even less knowledge of the biologic/genetic repercussions which might follow man's tampering;

88. Vance Packard, *The People Shapers*, (New York: Little, Brown, and Co., 1977), p. 11. (emphasis in original.)

89. Carl Rogers, cited in *Shapers, op. cit.*, Prefatory Comment.

90. Albert Rosenfeld, *The Second Genesis: The Coming Control of Life* (New York: Vintage Books, Random House Publishers, 1975).

91. Albert Rosenfela, "When Man Becomes As God: The Biological Prospect," *Saturday Review*, December 10, 1977, p. 15.

92. *Ibid.*

—that decisions as to whether to proceed, if at all, with genetic modification call for supremely skilled value judgments, a task for which traditional "value free" science is completely unequipped.

As to understanding past evolution, I submit that prolonged reliance on the traditionalist paradigm has left us more ignorant than not. It is open to question how long it will take to rectify the matter. As to present awareness of the potential implications of continued rushing to implement genetic technology — a haste unsupported by any decent justification — the clarity of Nicholas Wade's portrayal can hardly be improved. He writes, in *The Ultimate Experiment: Man-Made Evolution (DNA)*:

> "A turning point has been reached in the study of life
> . . . . Known at present by the awkward name of recombi-
> nant DNA research, the technique is in essence a method
> of chemically cutting and splicing DNA, the molecular
> material which the genes of living organisms are made of
> . . . What the new gene splicing technique [makes] possible
> is the transfer of genes from one species to another,
> regardless of the reproductive barriers that nature has built
> between them to isolate one species from another . . . The
> key to the living kingdom has been put into our hands . . .
>
> The ability to manipulate the stuff of life is an art of a
> different order, the ultimate technology. All technologies
> have their unintended and untoward side effects, and a
> technique of such power as gene splicing is unlikely to
> prove an exception . . . . The possible hazards of gene
> splicing all stem from the fact that there is as yet no predic-
> tive theory of evolution, no way of forecasting the effect of
> transferring genes from one species to another . . . The fact
> is that the technique involves playing evolution's game
> without fully understanding the rules. Until the rules are
> better understood, gene splicing inevitably carries the risk
> of making a forbidden play without knowing what the
> forfeit will be."[93]

To a very significant degree, man does not know *what* he is doing in grasping for control over evolution. Equally ominous, he has for-mulated no adequate answer as to *why* genetic change should be in-

93. Nicholas Wade, *The Ultimate Experiment: Man-Made Evolution (DNA)* (New York: Walker Publishing Co., 1977), pp. 1, 2, 4, 59, 146-7.

duced, if at all. Because of the vast confusion which exists concerning the relationship between levels and types of knowledge, especially between scientific and value-shaping theological and philosophical/metaphysical knowledge, he is stepping across the threshhold of evolutionary control ethically blindfolded. Words cannot describe the importance of getting the blindfold removed.

To emphasize the limitations of traditional science with respect to questions of value, the supreme example is that such science *cannot* answer the question: *Should* life on Earth be protected and perpetuated? If the answer to the question is "yes," then traditional science can provide endless help figuring out "how," but it is speechless before the most fundamental question of "should".[94]

Regarding values — their derivation, definition, and role — mankind is in a massive intellectual mess. The confusion began forming long ago with the basic assertions of Descartes, assertions which experience has totally discredited from an informed intellectual standpoint, but which have retained huge operative force.

I repeat. It was Descartes who gave serious promotion to the idea that only one type of knowledge and one means of verification is legitimate — the knowledge formulation and validation process derived from the natural sciences. Because metaphysical ideas, concerned with nontangibles such as context, ends, values, and purposes, do not and can not lend themselves to physical validation, metaphysical study was ruled worthless by definition. This "enlightened" dismissal of metaphysics has produced disastrous results. As Schumacher put it in *Small Is Beautiful:* "The leading ideas [mentioned earlier, all building on Cartesian assumptions] which claimed to do away with metaphysics, are themselves a bad, vicious, life-destroying type of metaphysics. We are suffering from them as from a fatal disease . . . It is not . . . true that metaphysics and ethics would be eliminated. On the contrary, all we got was bad metaphysics and appalling ethics."[95]

On top of that, most of society, thinking reductionist philosophy the only "modern" alternative, has been increasingly deferential to the viewpoint and outlook of constrained specialty disciplines. In *Guide,* Schumacher captures the essence of this calamitous state of affairs.

    "It is being loudly proclaimed *in the name of scientific*

---

94. Adler, *Conditions Of Philosphy, op. cit.,* especially Chapter 11.
95. Schumacher, *Small, op. cit.,* pp. 84-5.

objectivity that 'values and meanings are nothing but defense mechanisms and reaction formations', that man is 'nothing but a complex biochemical mechanism powered by a combustion system which energizes computers with prodigious storage facilities for retaining encoded information . . .'

"How is anyone to resist the pressure of such statements, made in the name of objective science unless, like Maurice Nicoll, he suddenly receives 'this inner revelation of knowing' that men who say such things, however learned they may be, *know nothing about anything that really matters?* People are asking for bread and they are being given stones . . . They long for guidance about how to live as responsible human beings, and they are told that they are machines, like computers, without free will and therefore without responsibility.

"'The present danger,' says Viktor E. Frankl, a psychiatrist of unshakable sanity, 'does not really live in the loss of universality on the part of the scientist, but rather in his pretence and claim of totality . . . What we have to deplore therefore is not so much the fact that *scientists are specializing,* but rather the fact that *specialists are generalizing.*' After many centuries of theological imperialism, we have now had three centuries of an ever more aggressive 'scientific imperialism,' and the result is a degree of bewilderment and disorientation, particularly among the young, which can at any moment lead to the collapse of our civilization. 'The true nihilism of today,' says Dr. Frankl, 'is reductionism . . . Contemporary nihilism no longer brandishes the word nothingness; today nihilism is camouflaged as *nothing-but-ness.* Human phenomena are thus turned into mere epiphenomena'."[96]

The combination of narrowly-oriented, specialist "intellectual" domination; the reduction of human phenomena to "nothing-but-ness"; the looming availability of the ultimate technology — biogenetic engineering of man — presents the utmost danger to present and future human life. Well before Packard, Leon Kass had been

---

96. Schumacher, *Guide, op. cit.,* pp. 5-6. (emphasis in original.) The behavior condemned by Frankl is the most dangerous possible type of ignorance-on-the-loose, scientific specialists totally unaware of the "mixed" nature of the situation and problems with which they are dealing.

warning — with insufficient effect — that

"We have paid some high prices for the technological conquest of nature, but none perhaps so high as the intellectual and spiritual costs as seeing nature as mere material for our manipulation, exploitation and transformation. With the power for biological engineering now gathering, there will be splendid new opportunities for a similar degradation of our view of man. Indeed, we are already witnessing the erosion of our idea of man as something splendid or divine, as a creature with freedom and dignity. And clearly, if we come to see ourselves as meat, then meat we shall become."[97]

Given the all-too-real potential for a repeat of Nazi-type experimentalism, for man again coming to be treated as mere meat, I perceive no aspect of paradigm overhaul more urgent than restoration of perspective regarding man. Let me here summarize what this essay has sought to accomplish, through creative synthesis, toward that end.

First, we have passed on historically unprecedented evidence that the universe had a start. We are, as certainly as intelligence and evidence can tell, participating in a Creation.

Second, we have brought to attention several demonstrations that the Creator can only be the "Original Author" of *Genesis.*

Third, through *Connection,* a century-plus of negative dogma has been dispelled and the Creator's positive purpose for both the Creation and human life (re)demonstrated.

Fourth, all available, scientifically valid evidence supports the belief that while man can be biologically related to several animal species for purposes of physiological comparison, he is unique, in a large number of telling respects. He is in fact a special creation, the most special of all, the consciously honored offspring of the Creator.

Just as life is not a simple stage in an evolutionary continuum, man is vastly more than a mere extension of animal life-forms. Each person's God-given status as a unique individual and his spiritual relation to the Creator demands the utmost respect. There is nothing in the presence of evolution, as an aspect of Creation, which should be allowed to impinge on that status or demand, save for the most

---

97. Leon Kass, "Making Babies — The New Biology and the 'Old' Morality," *The Public Interest,* Winter 1972, p. 53.

carefully considered exceptions. (I am completely cognizant of the fact that many existing and developing biomedical technologies have the potential for very positive contributions to human life and welfare. My comments pertaining to their application should not be read as a harsh and totally negative judgment, but as a strongly-worded effort to redress the severely imbalanced perception of transformists regarding the desirability or propriety of manipulation of human life.)

A global prohibition of scientific "free enterprise" manipulation of man must be solidly integrated into the rehabilitated world/life paradigm. The justification for the general prohibition is the foregoing four-point recapitulation. The preeminent concerns must be with the protection of the Creation and *intended* human life experience.

The moral problems inherent in the development of biomedical technologies will not be manageable without strong consensus as to who should serve "in the role of God". I argued in *Connection* that the always-to-exist *service* responsibilities of the Church be extended, that it assume and execute new responsibilities with respect to value guidance to govern ever-more-powerful scientific technologies, such as biomedicine. Rosenfeld reinforces my definition of the problem when he states: "The sociobiologists' emphasis on the genetic roots of human behavior raises fundamental questions about the human condition — questions that are philosophical, ethical, moral, and ultimately religious."[98]

Given the nature of the situation I must put an earnest question to those who will disagree (and there are certain to be many) with my general prescription. What other alternative is there for coping with the value guidance problem? It would be the height of moral irresponsibility to allow disagreement to be stretched into a denial of the problem or the need for an organized response to it. Critics cannot evade the most serious obligation of presenting a superior alternative.

The Church cannot possibly achieve contemporary and future success through reactive perpetuation of the world/life paradigm of Augustine and Constantine. The occurrence of the Second Vatican Council attests to recognition of that reality. I empathize with, and work to support, all who labor at responsible updating of the Church's basic perspective. At the same time, I fume at the defaulted promise of Vatican II regarding the specific issue at hand, the fault il-

---

98. Rosenfeld, *Saturday Review, op. cit.,* p. 16.

lustrated by Vatican reaction to birth of the first test-tube baby. It could but indulge in outrageously sophomoric cant: "interference with nature is not acceptable in any form". That equates positioning the Church, for all intent and purpose, against the use of *all* medicine during the course of a regular pregnancy. The standing doctrine of the Church on such matters is grievously lopsided, far out of touch with difficult reality.

It is, therefore, with grateful hope that we consider the reaction to that event of then Cardinal Albino Luciani, later Pope John Paul I: "I extend the warmest wishes to the English girl. As for the parents, I have no right to condemn them. Subjectively, if they acted in good faith and with good intentions, they could even gain merit before God . . . [But while] progress is a great thing . . . not all [change] is good for man. Will not science bear the appearance of the Sorcerer's Apprentice, who scatters mighty forces without, however, being able to dam or dominate them? Could there not be danger of a new industry arising, that of the manufacture of children? The individual conscience must always be followed, but the individual must make an effort to have a well-formed conscience. Conscience, indeed, does not have the duty of creating law, but of informing itself first on what the law of God dictates."[99]

If those remarks can be taken as an omen on behalf of strict but informed and sensitive balance, and translated into official, contemporary policy, the Church world/life paradigm will be moved enormously in the direction of restored credibility and attractiveness.

The work of overhaul must be pressed on all fronts. Outside the Church, two matters warrant particular attention. These have to do with clarifying relationships between types of knowledge and securing broad consensus on Purpose.

The earlier citations from Gilson give a firm insight into knowledge relationships. Still, with the long historical disparagement to which metaphysics has been subjected, I assume strong skepticism persists regarding its validity and utility. It is therefore exceedingly fortunate that a very up-to-date work is at hand which can banish such doubts and do so in the context of the foundation of natural science — physics. I refer to *The Road of Science and The Ways To God* by Stanley L. Jaki.[100] Why is it, Jaki asks, that the truly penetrating scientific discoveries about the natural world and the universe have

---

99. *Time,* September 4, 1978, p. 66.

100. Stanely L. Jaki, *The Road of Science and The Ways To God,* (Gifford Lectures), (Chicago: The University of Chicago Press, 1978).

taken place within only the Judeo-Christian cosmological frame of reference? In the course of answering that question, Jaki shows explicitly the commonality of the worldview of the great scientific contributors from Copernicus through Einstein and the necessary linkage between sound metaphysics and scientific advance. Jaki is another individual concerned with clearing away the philosophical debris of materialistic scientism. Toward that end he has "placed certain papers on the table," in effect setting an insuperable challenge before any who would deny the role of metaphysics in valid human understanding.

We must restore a proper alignment of the various forms of knowledge to assist in appropriate control over science and technology. We must also secure a broad consensus on ultimate Purpose. I doubt seriously that a merely general reconciliation between science and religion will suffice (the type which Rosenfeld suggests has already begun). I've sought to explain why in Chapter V of *Connection,* from which the following excerpt is taken. It is a portion of the explanation as to why the long-held goal of earthly Utopia is neither possible nor desirable, and why Tension will remain a central feature of man's personal and social experience.

—The ever-present danger of conflict between secular and higher law. Caesar will face an unrelenting succession of problems — and the citizens must support his responsible efforts, give strength to his considered innovations. We must render unto Caesar all his due — but that does not mean our everything, in particular our final allegiance. The interaction between church and state, since it revolves around the common subject and interest of people, cannot be avoided. Yet the concerns and responsibilities of the two institutions are not identical and are merged only at the serious risk of each other's essential interests. Consequently, it is necessary for an American-type wall between church and state, no matter how blurred, indistinct, and variable a barrier it will prove to maintain, to be built into every portion of the human camp.

—The boomerang tension of knowledge. To our chagrin, and in refutation of the Enlightened assumption that intensive investigation and effort will produce ultimate insights, we have learned that *while knowledge advances, ignorance also grows.* As the Creation matures, questions and problems multiply. The more we learn, the more we need to learn. No attitude but humility is proper in comparing what we know compared to what might be known, and in seeing that man's "solutions" are constantly producing new problems. Social advance carries in it the seeds of social disunion; inconsiderate exploitation of "free" goods such as air and water tends to make their maintenance ever more expensive; use of medicines and drugs which assist in curing one illness but which in the same process promote genetic development of micro-organisms more resistent to medicines and drugs; and so on . . .

—As a corollary, we now know that *new is not synonymous with better*. We have learned that, though technological and other change or novelty may be possible, *the possible may be senseless*. To change things for the sake of change, to do things simply because we have the technological or other wherewithal to do them, is to risk social and ecological disaster.

—Further, finally, complicating the situation, calculating the costs and benefits of the possible *cannot* be done without reference to overriding objectives — overriding purpose. And the definition of purpose is not susceptible to quantitative or mechanical formulation. No computer can handle the job. No machine can relieve the human intellect of this most fundamental responsibility. Without an intellectually sound and respectable, widely shared, governing purpose to channel, direct, *and permit evaluation* of social activity, man — particularly technological man — could, and likely would, destroy both the Creation and himself. *To repeat, to underscore: definition of Purpose is the key to both human and humane survival.* [101]

There is not an iota of evidence that human nature has changed or is naturally changing since it appeared in its intended (created) form thousands or tens of thousands of years ago. Human maturation, individual and social development, has occurred but such change has taken place within designed limits. A "set piece" has been maturing. There has been no evolutionary change in the piece we know as human nature since it was first set. Yet the biomedical transformists,

---

101. Hadd, *Connection, op. cit.,* pp. 119-120. (emphasis in original.)

as large an example of the Sorcerer's Apprentice as can be found, could induce wholesale change.

Biomedicine poses three types of dangers. The first, physical, has already been explained in the quotation from Wade; means are becoming available to dismantle, rearrange and reassemble basic components of living organisms, including the human. The second and third kinds of danger are social and spiritual. We must not permit biomedical modifications unless and until the following kinds of questions are given acceptable answers. If you would undertake to change the long-set constitution of man, why? And if change, to what extent? How do you define "progress" and "improvement" in man's makeup, given God's design for human individuality coming to fruition in diversified social community? Would the transformers design away "normal distribution" on behalf of arbitrarily determined, genetically monotonous stock just to satisfy *their* preferences? By what criteria are potentially inducible changes in the natural composition of society to be judged?

The spiritual danger is the worst of all. Biogenetic manipulation can lead to a direct, total affront to God. He has set forth a Creation intended to serve as a training ground for individual human souls. Biomedicine, specifically the procedure of cloning, has the potential for creating soulless forms of "human" life.[102] I would think the sentence upon Judas mild by comparison to the wrath to be encountered by any person contributing to the production of soulless forms of man.

David Rorvik claims in his book, *In His Image: The Cloning Of A Man,* that the first such event has already occurred.[103] His "report" has been judged as, at best, a sophisticated hoax.[104] Some scientists doubt that human cloning can ever be achieved. Among the majority of those closest to the scene, however, the primary doubt appears to be not "can it be done?" but "has it been done?". Packard's survey of biotechnological developments results in the judgment that "Certainly, unless human cloning is banned, it is likely to become a reality within the lifetime of the great majority of people now alive."[105]

Rorvik, displaying the teleological blindness of materialistic scientism, delights at the prospect of human cloning. ". . . one can put the puzzle together and proceed to clone. What is needed is the will to do

102. Rosenfeld, *Second Genesis, op. cit.,* pp. 91-94.
103. David Rorvik, *In His Image: The Cloning of a Man* (New York: J. B. Lippincott Publishers, 1978).
104. Leonard Isaacs, "The Once and Future Clone," *Hastings Center Report,* June, 1978.
105. Packard, *Shapers, op. cit.,* p. 269.

this and the money, resources, and appropriate medical team. The desire to do it is the most important thing . . ."[106] At least Rorvik makes unmistakably clear the central importance of perspective and motivation relative to applications of biomedicine. It is *not* inevitable that the human future, in biomedical terms, will be "quite different" from the past or present. If it is different it will be the result of individual and social choice. No amorphous "biology" is going to dictate the shape of the future. There is *no* technological "imperative" to change.

Let us establish firm rules to control applications of biomedicine, henceforth always in the light of a fully defensible teleological definition of the purpose of human life. Given it, the burden of positive proof is set completely on would-be "transformers". I submit that past and present means for calculating social and scientific "advancement" and "benefit" are pathetically inadequate, and more so than ever when it comes to rationales on behalf of manipulation of the genetic nerves of the Creation and its capstone, soul-graced *Homo sapiens.*

The family has never been confronted with so momentous an issue: "let's see what man can make of man". Never has the need for *wise* choice been so imperative. The Weinbergs of the scientific community — who sense at heart that things are badly out of balance — and the public, which is funding the majority of biomedical research, are going to have to apply themselves with dedication to prevent the situation from getting worse.

We must promote hearing time for, and somber heeding of, the Adlers, Gilsons, Jakis, and Kasses and be prepared to apply the competent guidance expected from morally *informed* secular quarters as well as the Church.

* * * * *

In return for the privilege of experiencing conscious, aware earthly life, man inherits certain obligations. One of his very largest is stewardship of the Creation, in order that it might function for an indefinite period as basically designed. (The prospect of universal "heat death" pertains to a highly distant future, so far beyond the present as to have no practical relevance for present generations and eons of them to come.)

Man, collectively, is betraying the obligation of stewardship. Under the influence of materialistic scientism, he has come to view

---

106. David Rorvik, interview comment reported in *The Washington Star,* March 30, 1978.

nature as "mere material for manipulation, exploitation, and transformation" and, typically, for immediate benefit with no concern for the morrow, especially more distant ones. This aspect of perspective is identical to all existing ideological systems — capitalist, communist, socialist.

Furthermore, man's secular actions, running roughshod over nature, have long been abetted by certain invalid Christian attitudes and ideas about man's "dominion" over nature. The command given in *Genesis* for man to "subdue" the Earth has been badly misinterpreted. Properly translated, the Hebrew word "subdue" connotes not so much unbridled privilege as sober responsibility.[107] Man's task is to cultivate the Earth, dress and keep it in accord with natural (designed) constraints, and not engage in selfish, short-term destructive gouging and exploitation of natural wealth.

Lester R. Brown, director of Worldwatch Institute, has documented the most significant ways in which the Creation is being mutilated. He has shown that the four principal biological systems on which humanity depends — fisheries, forests, grasslands, and croplands — are being badly overtaxed and, without signficant change in man's management approach to them, an ecological disaster of massive proportions will be upon us.[108] The trends of man-induced systemic deterioration must be reversed, and we must also move decisively in the development of renewable energy resources. Sustaining the Creation is a most serious business and man must cease abrogating his responsibilities. A reminder, a thought from the Koran, is presently very much in order:

"The heaven and the earth and all between, thinkest thou I made them *in jest*?"[109]

It is going to take a highly concerted effort and substantial resources to reverse and offset the damage being done to the Spaceship Earth's life-supporting ecosystems. Those resources are readily available, provided the unjustifiable ideological confrontation between East and West is ended. The confrontation, fueled by false science, sustains the insane global preoccupation with armaments spending mentioned at the outset of this book. That preoccupation, in turn, detracts directly from adequate attention and

---

107. See Bales, *Genesis Account, op. cit.,* p. 47.

108. Lester R. Brown, *The Twenty Ninth Day* (New York: W. W. Norton and Co., 1978).

109. The Koran, as cited by C. S. Lewis in *God In The Dock* (Grand Rapids, Michigan: Eerdmans Publishing Company, 1970), p. 182.

resources being applied to the real problems of the era, such as ecosystem destruction. Worldwide military spending now exceeds $750,000 per minute, $45 million per hour, $400 billion annually. The elimination of false science — leading to annulment of nuclear force and other armaments strategizing by all members of the family — will allow positive deployment of available resources and allow us to get to work, in safety, on ecosystem and social problems either threatening or already impinging on the quality of human life and the prospects for a decent future.

There is no nation, or major institution, which can consider itself exempt from the life-preserving, Creation-sustaining demands that are bound to follow from the dictates of a properly overhauled world/life paradigm. There can be no exemptions to changed behavior, but neither any superficial moral posturing. No one is in a legitimate position to "cast the first stone".

Like every nation state, the Church must face up to moral responsibility. It *must* reform its doctrine pertaining to birth control and take seriously the lessons of nature regarding "carrying capacity". Pressures on the ecological life-support facilities of Spaceship Earth worsen daily. This is due, in part, to use of faulty technologies; but at base it is a remorsefully burgeoning world population which "drives the system" and leaves in its wake ever more environmental disruption, resource depletion, and social degradation.

Scientists are not omniscient. Politicians are not magicians. There are physical limits to the amount of economic wealth any country can produce, and corresponding limits to the number of persons who can be provided for decently at any given time. There is no realistic chance of decently accommodating on Spaceship Earth all the persons who might be born in one particular time-frame. Population numbers must be controlled, else we will have a theologically-supported, if not induced, Hell-on-Earth. If the Church is really concerned about "the least of these" it will no longer delay rectifying doctrine which is working to their direct disadvantage.

* ** ** ** ** ** *

The Bible teaches that history is a *sacred* process which takes place, through the Creation, for the sole benefit of man, and at the greatest possible personal expense to God. The present generations have an obligation to restore the place of those teachings to the center of individual and collective life. The task of properly overhauling the world/life paradigm is a task of highest urgency and importance. Let us pursue it, and the ensuing work, diligently and with the attitude both suggested and lived out by one of this era's finest mentors, Ernst Friedrich Schumacher.

> What we need are what I would call optimistic pessimists who can see clearly that we can't continue as before, but who have enough vigor and joyfulness to say, all right, so we change course.[110]

If we heed that advice, then long-term protection of the Creation will be achieved and we will have proven ourselves worthy offspring of the God of purposeful Life.

---

110. Schumacher, cited in *The Futurist* (Journal of the World Future Society), December 1974, p. 284.

INDEX

Macbeth, Norman 42, 44, 65
Marx, Karl 7, 8, 10n
McGinley, Phyllis 22n
metaphysics 8, 74, 78

natural selection 8, 17, 18, 32,
   42-3, 45
nuclear arms race 6, 7, 8, 82-3
nuclear "fail-safe" control
   systems 6

overpopulation 1, 84

Packard, Vance 71, 72n
paradigm, world/life
  —conflicting perceptions 5, 7,
    45
  —contribution of religion 10,
    60
  —definition 5
  —of the Church 77, 84
Pasteur, Louis 30
Penzias, Arno H. 28
Pilbeam, David 15n, 54
purpose, ultimate 24, 55, 60
   79-80, 82-85

Rodgers, Carl 72
Rorvik, David 81, 82
Rosenfeld, Albert 72, 77, 79

Schof, J. William 15n
Schumacher, E.F. 7, 9, 10, 11,
   34, 60, 65, 74, 85
sociobiology 15, 54n, 77
Steady State cosmology 28
Stern, Karl 64
stewardship responsibilities 82-4
Strategic Arms Limitation
   Treaty 6
Sutherland, John W. 23n

Velikovsky, Immanuel 25n, 66

Wade, Nicholas 73, 81
Washburn, S.L. 54n, 55n
Weinberg, Steven 27, 56
Wilson, Edward O. 15
Wysong, R.L. 31-6, 42, 44,
   51n, 64, 65

ACKNOWLEDGMENTS

Grateful acknowledgment is given to the following publishers for permission to quote from their publications.

DODD, MEAD & COMPANY
The Everlasting Man, by G.K. Chesterton (Apollo edition, 1971).

DEHOFF PUBLICATIONS
The Transformist Illusion, by Douglas Dewar.

HARPER & ROW, PUBLISHERS, INC.
Small Is Beautiful: Economics As If People Mattered, by E.F. Schumacher, (New York; Harper & Row, 1973). Copyright © 1973 by E.F. Schumacher. Reprinted by permission of the publisher.

A Guide For The Perplexed, by E.F. Schumacher, (New York: Harper & Row, 1977). Copyright © 1977 by E.F. Schumacher. Reprinted by permission of the publisher.

INQUIRY PRESS
The Creation-Evolution Controversy, by R.L. Wysong.

NATURAL HISTORY MAGAZINE
"This View Of Life," by Stephen Jay Gould, copyright © The American Museum of Natural History. (Reprinted in *Ever Since Darwin: Reflections In Natural History,* by Stephen Jay Gould (New York: W.W. Norton and Co., 1977).)

YALE UNIVERSITY PRESS
God and Philosophy, by Etienne Gilson.

A final note. Sympathizers, like myself, of responsible women's liberation will, I hope, forgive my use of 'man' and 'his' to cover both sexes. As has been well-said elsewhere, "to identify both every time is otiose; to write constantly of 'persons' seems clumsy. Personkind has troubles enough."